T0185653

Lecture Notes in Electrical Engineering

Volume 663

The book series *Lecture Notes in Electrical Engineering* (LNEE) publishes the latest developments in Electrical Engineering—quickly, informally and in high quality. While original research reported in proceedings and monographs has traditionally formed the core of LNEE, we also encourage authors to submit books devoted to supporting student education and professional training in the various fields and applications areas of electrical engineering. The series cover classical and emerging topics concerning:

- Communication Engineering, Information Theory and Networks
- Electronics Engineering and Microelectronics
- Signal, Image and Speech Processing
- Wireless and Mobile Communication
- Circuits and Systems
- Energy Systems, Power Electronics and Electrical Machines
- Electro-optical Engineering
- Instrumentation Engineering
- Avionics Engineering
- Control Systems
- Internet-of-Things and Cybersecurity
- Biomedical Devices, MEMS and NEMS

For general information about this book series, comments or suggestions, please contact leontina.dicecco@springer.com.

To submit a proposal or request further information, please contact the Publishing Editor in your country:

China

Jasmine Dou, Associate Editor (jasmine.dou@springer.com)

India, Japan, Rest of Asia

Swati Meherishi, Executive Editor (Swati.Meherishi@springer.com)

Southeast Asia, Australia, New Zealand

Ramesh Nath Premnath, Editor (ramesh.premnath@springernature.com)

USA, Canada:

Michael Luby, Senior Editor (michael.luby@springer.com)

All other Countries:

Leontina Di Cecco, Senior Editor (leontina.dicecco@springer.com)

** Indexing: The books of this series are submitted to ISI Proceedings, EI-Compendex, SCOPUS, MetaPress, Web of Science and Springerlink **

More information about this series at http://www.springer.com/series/7818

Andrea Cataldo · Nicola Giaquinto ·
Egidio De Benedetto · Antonio Masciullo ·
Giuseppe Cannazza · Ilaria Lorenzo ·
Jacopo Nicolazzo · Maria Teresa Meo ·
Alessandro De Monte · Gianluca Parisi ·
Federico Gaetani

Basic Theory and Laboratory Experiments in Measurement and Instrumentation

A Practice-Oriented Guide

 Springer

Andrea Cataldo
Department of Engineering for Innovation
University of Salento
Lecce, Italy

Egidio De Benedetto
Department of Engineering for Innovation
University of Salento
Lecce, Italy

Giuseppe Cannazza
Department of Engineering for Innovation
University of Salento
Lecce, Italy

Jacopo Nicolazzo
Department of Engineering for Innovation
University of Salento
Lecce, Italy

Alessandro De Monte
Department of Engineering for Innovation
University of Salento
Lecce, Italy

Federico Gaetani
Department of Engineering for Innovation
University of Salento
Lecce, Italy

Nicola Giaquinto
Department of Electrical and Information
Engineering
Polytechnic University of Bari
Bari, Italy

Antonio Masciullo
Department of Engineering for Innovation
University of Salento
Lecce, Italy

Ilaria Lorenzo
Department of Engineering for Innovation
University of Salento
Lecce, Italy

Maria Teresa Meo
Department of Engineering for Innovation
University of Salento
Lecce, Italy

Gianluca Parisi
Department of Engineering for Innovation
University of Salento
Lecce, Italy

ISSN 1876-1100   ISSN 1876-1119   (electronic)
Lecture Notes in Electrical Engineering
ISBN 978-3-030-46742-5   ISBN 978-3-030-46740-1   (eBook)
https://doi.org/10.1007/978-3-030-46740-1

This Springer imprint is published by the registered company Springer Nature Switzerland AG
The registered company address is: Gewerbestrasse 11, 6330 Cham, Switzerland

Foreword

Anyone who has ever taught either Human Sciences and Scientific subjects knows how important it is for an effective learning to let students experience first-hand what they have studied from theory.

Especially with scientific subjects, what you experience in practice does not always perfectly collimate with what you expect from theory, especially when non-idealities come into place. For example, in my courses of Electric and Electronic Measurements, most of my students do not really appreciate how the variability in the indications of a measurement device affects the uncertainty or the impact of noise on signals, until they're faced with the practical effects of these phenomena.

Generally, it is very hard to find the optimal trade-off between theory and practice. This is of pivotal importance, as courses have limited duration, and any dedicated professor wants to make to deliver good lessons. In spite of all the best intentions, often, there is not enough time to do this. Through experience, I think I have found an optimal balance of theory and practice in my courses. However, when Andrea Cataldo talked to me about this book project, I really welcomed his idea for two main reasons.

The first is that the book places itself as a sort of bridge between theory and practice. The book proposes a series of laboratory experiences, all relevant to the electric and electronic measurements, with a limited but sufficient theoretical details to accomplish two tasks. On one hand, the book represents a useful and practical guide for self-learning students or practitioners. On the other hand, this book could serve as a good starting point for teachers and professors who, with limited effort, can adapt the proposed laboratory experiences to the specific needs of their courses. As a member of the Metrology Community, I also find of the utmost importance that this book rigorously covers all the metrological aspects of measurements, while still preserving a hands-on approach.

The second reason why I praise the idea of the idea of this book is an ethical one. In fact, I know that the core of this book was conceived from laboratory reports of dedicated students, who are all duly acknowledged as authors.

I believe that this conveys a very positive message to students, who understand that they are not "passive recipients of knowledge", but they are major players in the process. And this is something that resonates with my professional and personal beliefs.

Naples, Italy   Leopoldo Angrisani
February 2020

Preface

The idea of this book started while I was assessing the laboratory reports of my students from the course of "Measurements for Telecommunications" of the MS Degree in Telecommunication Engineering. The reports submitted by the students were detailed, contextualized and suitably commented on. They showed that the students had put not only effort but also immense care in drafting those reports. I like to think that they liked the course so much that they wanted to "transfer" what they learned. I admit that that was probably my biased perception of their work; nonetheless, I thought that their effort and dedication could inspire their fellow students to study with as much passion and could help their peers in approaching measurements. This is what triggered me to plan this book.

Indeed, it is very common of lecturers and professors to ask their students to provide reports or presentations which are later re-adapted, "filtered" by the lecturers for their own educational purposes. However, I did not want this to be the case; and, in fact, the present book overturns this approach. I decided to start an experiment of a constructive interaction between students and teacher, where the students could be the "lead actors" in a project in which they could share with their peers (and not only) the results of their work. This book is the result of this process.

As you will read, each chapter (from 2 to 4) proposes a number of practical laboratory experiments and the reports (including pictures, images and figures) of actual laboratory experiments. The topics of the laboratory experiments relate to fundamental measurement cases such as those typically performed in the time and frequency domains.

Some brief theoretical considerations are also included in each chapter to support the laboratory experiences. The theoretical considerations are certainly not exhaustive and they are not intended to. In fact, this book must not to replace well-established theory textbooks. On the contrary, the book is meant to represent a "supporting companion" to those books. In my experience as a professor, I have always felt that this has been missing.

That said, one should not fall either in the error of dismissing the book as a mere list of practical experiments. In fact, all the laboratory experiences are accompanied by a metrological characterization of the measurements experiments. This is why I invited a colleague of mine, prof. Nicola Giaquinto, to contribute one chapter of his own to the book (Chap. 1), which provides a concise but rigorous and robust description of measurement uncertainty theory. These very important theoretical considerations on measurements uncertainty, summarized by my friend Nicola, are then applied to the practical laboratory measurements. The final result is a book that, in my opinion, successfully merges two contrasting souls: practical measurements and the related basic theoretical references with a particular care to the related uncertainty evaluation.

I hope that this book will represent a useful support for both teachers and students in the field of electric and electronic measurements.

Lecce, Italy   Andrea Cataldo
February 2020

The original version of the book was revised: The co-author's name has been corrected in the book. The correction to the book is available at https://doi.org/10.1007/978-3-030-46740-1_5

Contents

Acronyms

ADC   Analog to Digital Converter
AM   Amplitude Modulation
CMRR   Common mode rejection ratio
CRT   Cathode Ray Tube
DFT   Discrete Fourier Transform
DPO   Digital Phosphor Oscilloscopes
DSO   Digital Storage Oscilloscopes
FFT   Fast Fourier Transform
FM   Frequency Modulation
FSK   Frequency Shift Keying
GBP   Gain Bandwidth Product
GUF   GUM Uncertainty Framework
GUM   Guide to the expression of uncertainty in measurement
IF   Intermediate Frequency
LPE   Law of Propagation of Errors
LPU   Law of Propagation of Uncertainty
LSB   Least Significant Bit
MSO   Mixed Signal Oscilloscopes
PCB   Printed Circuit Board
PSD   Power Spectral Density
RBW   Resolution Bandwidth
RF   Radio Frequency
SA   Spectrum Analyzer
SR   Slew Rate
SWR   Standing Wave Ratio
TDR   Time-Domain Reflectometry
THD   Total Harmonic Distortion
VBW   Video Bandwidth

VCO	Voltage Controlled Oscillator
VNA	Vector Network Analyzer
VSWR	Voltage standing wave ratio
WCU	Worst-Case Uncertainty

Chapter 1
Basic Theory of Uncertainty Evaluation in Measurements

Abstract This chapter illustrates basic concepts necessary to justify and understand the uncertainty evaluations presented throughout this book. The aim is to provide a brief and practically useful explanation of fundamental concepts and equations, not a complete theory of measurement uncertainty (which could easily be the subject of an entire book). The main source for the symbols, terminology, and concepts used in this chapter is the authoritative document "Guide to the Expression of Uncertainty in Measurement"(GUM). The reader is also encouraged to refer to the "International Vocabulary of Metrology" (VIM), which has a more extensive discussion of terminology, sometimes with slight differences with respect to the GUM. Besides the basic elements of the GUM theory, this chapter illustrates how to handle uncertainties due to gain, offset, nonlinearity, and quantization errors. Such knowledge is necessary for understanding the accuracy specifications of many real-world instruments. The chapter concludes with examples, accompanied by relevant explanations, of accuracy specifications of actual instruments.

1.1 Introducing Uncertainty in Instruments: A Real-World Example

If we measure a temperature with a commercial thermometer, how accurate is the result we get? The answer can usually be found in the instrument's specifications. Let us consider, for example, the Fluke 62 MAX Mini Infrared Thermometer. An excerpt from the specifications [1] is shown in Fig. 1.1.

Based on these figures, if our measurement is, say, 20.0 °C, the "accuracy" is ± 1.5 °C, since it is the greatest number between 1.5 °C and 1.5% of the reading = (1.5/100) × 20 = 0.3 °C. The intuitive meaning of this result is quite clear: we can ensure, with a high level of confidence, that the measurement error is in the range ±1.5 °C, or in other words, we can ensure with a high level of confidence that the "true" temperature is in the interval [18.5, 21.5]. If we have measured the

Specifications	
Temperature range	-30°C to 500°C (-22°F to 932°F)
Accuracy	±1.5°C or ±1.5% of reading, whichever is greater
	-10°C to 0°C: ±2.0
	-30°C to -10°C: ±3.0
Response time (95%)	< 500 ms (95% of reading)

Fig. 1.1 Example of uncertainty specifications in a measuring instrument (Fluke 62 MAX Mini Infrared Thermometer [1])

temperature for a specific purpose, e.g., for controlling an industrial process, this is important information, at least as important as the measurement result itself.

"Accuracy" is a preferred term in commercial instrument specifications, since it carries a positive meaning. The term preferred in the scientific literature and in technical standards is "uncertainty." "Accuracy" is not avoided or deprecated, but it is reserved to express a qualitative concept. An instrument may be more or less accurate, but the accuracy is quantified in terms of measurement uncertainty.

In the following, we illustrate some basic concepts necessary to justify and understand the uncertainty evaluations presented throughout this book. Our main source for symbols and terminology will be the authoritative document "Guide to the expression of uncertainty in measurement" (GUM) [2, 3]. For example, accuracy is defined as "closeness of the agreement between the result of a measurement and a true value of the measurand" in GUM B.2.14, which specifies that it is "a qualitative concept." It is also useful to consult the "International Vocabulary of Metrology" (VIM) [4, 5], which has a more extensive discussion of the terminology, sometimes with slight differences with respect to the GUM.

1.2   Measurand, Value of the Measurand, Measurement Results

Measuring is a process, called *measurement procedure* (GUM, B.2.8), involving a *measurand*, i.e., the quantity of interest (GUM, B.2.9), and a *measurement result*, i.e., the value that the procedure attributes to the measurand (GUM, B.2.11). The relationship between measurand and measurement result is quite obvious, and is represented in Fig. 1.2.

It must be noted that according to the GUM, the word "measurand" denotes the physical quantity subject to measurement (like "the temperature of this point of this surface"), and not its *value*. The *value of a quantity* is a number, multiplied by a unit of measurement (GUM, B.2.2), expressing the magnitude of the quantity. Numbers and units of measurement are of course entities of different nature: therefore the multiplication has a formal meaning, and is represented as a juxtaposition. The *value*

Fig. 1.2 Quantities in a measurement procedure

of the measurand and the measurement results are therefore both numbers, multiplied by the same unit of measurement. They are denoted, in Fig. 1.2 and below, by the symbols x and y respectively.

A few words are useful about the definition of the value of the measurand given by GUM B.2.2. It is a value "consistent with the definition" of the measurand, and can be seen as the measurement result "that would be obtained by a perfect measurement." In this respect, x can also be called "true value" (of the quantity), as opposed to y, which is the "measured value." The expression "true value" is recognized and admissible (GUM, B.2.3), but in the expression "true value of the measurand," the word "true" is considered logically redundant. Indeed, two different words are used to refer to the input of the measurement procedure, although one is enough. To summarize:

1. value of the measurand = true value = true value of the measurand = x;
2. measurement result = measured value = y.

The value of the measurand x is, by definition, unknown. The aim of the measurement procedure is to get as good an approximation y of that value as possible. Sometimes, however, it is necessary to attach to x an actual value, separate from y. Any actual value given to x is called a "conventional true value" (GUM, B.2.4) or "reference value" (for this expression, see also VIM, 5.18). For example, if one wants to test the metrological performance of a thermometer, the same temperature must be measured with it, the "instrument under test," and with another thermometer of much better accuracy, the "reference instrument." The instrument under test gives a measured value y, while the reference instrument gives a "conventional true value" or "reference value" x.

Finally, we note that the GUM does not consider the true value unique: "there may be many values consistent with the definition of a given particular quantity" (GUM, B.2.3). On the other hand, the GUM itself, in its first section (GUM, 1.2), states that the measurand "can be characterized by an essentially unique value." These seemingly contrasting statements have, however, a very logical interpretation. We must think of x as an *undetermined unique value*, but we must not be surprised if *a number of different values can be conventionally assigned to x*. In other words:

- saying "the conventional true value" is deprecated: we should always say "a conventional true value";
- saying "the true value" to refer to an undetermined quantity, and not to a specific value, is acceptable, even if we may prefer "the value of the measurand."

The word "true" is of course problematic from an epistemological viewpoint, but it must be used only to refer to x in Fig. 1.2.

1.3 Measurement Error

1.3.1 Definition and Statistical Nature of the Measurement Error

With the symbols x for the value of the measurand and y for the measurement result, the *measurement error* is given by (GUM, B.2.19)

$$e = y - x . \tag{1.1}$$

The error is, of course, unknown: if it were known, one could obtain the true value with the trivial operation $x = y - e$.

Since the error is unknown, in order to quantify its magnitude, i.e., to evaluate the measurement uncertainty, it is natural to use statistical concepts. In the following, we will assume that the reader is aware of the basic concepts of probability theory and is familiar with the main probability distributions and related computations.

From the knowledge of the measurement result y and from information about the measurement procedure, it is possible *to associate to the error a probability density function* (pdf) $f_e(\epsilon)$. This pdf is a "probability distribution characterized by the measurement result" (an expression in the GUM, in 2.3.5, 6.3.2, G.1.3 etc.).

For reasons of simplicity and clarity, we will define the measurement uncertainty *on the basis of the pdf of the measurement error*. It must be noted that the GUM defines the uncertainty in a more involved way, but throughout it, the concept of error is ubiquitous, and it is specifically stated that (GUM, 2.2.4):

> The definition of uncertainty of measurement … is an operational one that focuses on the measurement result and its evaluated uncertainty. However, it is not inconsistent with other concepts of uncertainty of measurement, such as a measure of the possible error in the estimated value of the measurand as provided by the result of a measurement.

Before proceeding to theoretical considerations, definitions, and equations, it is useful to provide two practical examples of error pdfs.

1.3.2 Measurement Error with Normal (Gaussian) pdf

Suppose that the measurand value is an unknown constant voltage $x(t) = x$, and that the measurement result y results from taking a sample of the signal $y(t) = x(t) + n(t)$, where $n(t)$ is electronic thermal noise with known RMS value $V_{RMS} = 0.025$ V. From the physics of the thermal noise, the pdf of the error $e = y - x$ is normal (or Gaussian), with mean $\mu = 0$ and standard deviation $\sigma = V_{RMS}$:

$$f_e(\epsilon) = \frac{1}{\sigma \sqrt{2\pi}} \exp\left[-\frac{1}{2} \frac{(\epsilon - \mu)^2}{\sigma^2} \right] . \tag{1.2}$$

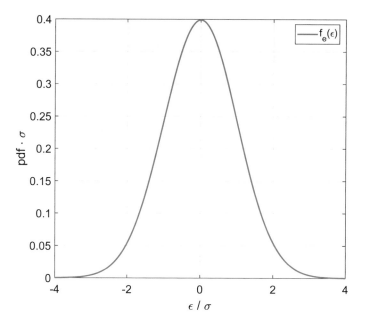

Fig. 1.3 Example of normal pdf of measurement error (thermal noise with known RMS value)

To express that the error has this pdf, we can write

$$e \sim N(\mu, \sigma^2),   \tag{1.3}$$

with $\mu = 0$, $\sigma = V_{RMS}$. The zero-mean normal distribution is depicted in Fig. 1.3.

1.3.2.1 Computations of Symmetric Intervals in the Case of Normal pdf

In many applications, it is necessary to calculate the probability that a normally distributed variable lies in a certain interval. In the context of the GUM Uncertainty Framework, or GUF [6], it is especially important the quantity

$$p = \text{Prob}(-k \cdot \sigma \le e \le k \cdot \sigma)   \tag{1.4}$$

i.e., the probability that the normal zero-mean error $e \sim N(\mu, \sigma^2)$ lies in the symmetric interval $[-k \cdot \sigma, k \cdot \sigma]$. The relations linking k and p are well known from basic probability theory. For a given k, p is given by

$$p = 2 \cdot \Phi(k) - 1,   \tag{1.5}$$

Table 1.1 Commonly used values of k, and corresponding p

k	p (%)
1	68.27
2	95.45
3	99.73
4	99.99

Table 1.2 Commonly used values of p, and corresponding k

p (%)	k
90	1.64
95	1.96
99	2.58
99.9	3.29

and conversely, for a given p, k is given by

$$k = \Phi^{-1}\left(\frac{1+p}{2}\right).$$ (1.6)

The function $\Phi(z)$ denotes the *standard normal distribution*, i.e., the cumulative distribution function (cdf) of the standard normal variable $X \sim N(\mu, \sigma^2)$ with $\mu = 0$, $\sigma^2 = 1$; the function $\Phi^{-1}(\alpha)$ denotes its inverse, the standard quantile function. Values of k and p commonly used in metrology appear in Tables 1.1 and 1.2.

In metrology, the association of the value $k \cong 2$ with the probability $p \cong 95\%$ is particularly common.

1.3.3   Measurement Error with Uniform pdf

Suppose that a measurement result, e.g., $y = 10.2$, is obtained by roundoff quantization of the unknown true value x, with a quantization step $Q = 0.1$. For $y = 10.2$, the true value must be in the interval $\left[y - \frac{Q}{2}, y + \frac{Q}{2}\right] = [10.15, 10.25]$. The error $e = y - x$ must lie, in general, in the interval $\left[-\frac{Q}{2}, \frac{Q}{2}\right] = [-0.05, 0.05]$. Without further information, there is no reason to consider any value of the error more probable than any other. The pdf of the error is therefore uniform in $[-U_{max}, U_{max}]$, with $U_{max} = \frac{Q}{2}$:

$$f_e(\epsilon) = \begin{cases} \frac{1}{2U_{max}} & |\epsilon| \leq U_{max}, \\ 0 & |\epsilon| > U_{max}. \end{cases}$$ (1.7)

A graph of this pdf is shown in Fig. 1.4.

To express that the error has this pdf, we can write

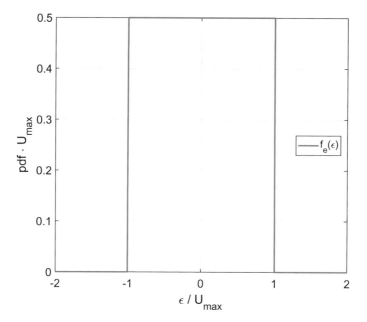

Fig. 1.4 Example of uniform pdf of measurement error (roundoff quantization)

$$e \sim \text{Unif}\left(-U_{max}, U_{max}\right).\tag{1.8}$$

1.4 Measurement Uncertainty

1.4.1 General Definition

According to the GUM, measurement uncertainty is a "parameter, associated with the result of a measurement, that characterizes the dispersion of the values that could reasonably be attributed to the measurand" (GUM, B.2.18). The uncertainty of the measurement result y is in general denoted by $U(y)$, with the uppercase letter U, for "uncertainty." The notation U_y is also used. If $U(y)$ is the uncertainty of the measurement result y obtained by applying a measurement procedure to a measurand with unknown value x (see Fig. 1.2), we will write (GUM, 6.2.1)

$$x = y \pm U(y).\tag{1.9}$$

An equivalent expression, allowed by the GUM, is

$$y - U(y) \le x \le y + U(y).\tag{1.10}$$

It is important to note that:

- in these expressions, y is meant to be the best estimate of x;
- quantifying the uncertainty of y using a *single parameter $U(y)$* is allowed by the hypothesis of a symmetric pdf. This is an underlying hypothesis of the GUM Uncertainty Framework [6], and will always be assumed in this book.

1.4.2 "Type A" and "Type B" Evaluation of Uncertainty

Before explaining how the parameter $U(y)$ is derived by the error pdf, it is necessary to clarify an important distinction, that between "type A" and "type B" evaluation of uncertainty.

According to the GUM, the type A evaluation is obtained "by the statistical analysis of series of observations" (GUM, 2.3.2), while the type B arises "by means other than the statistical analysis of series of observations" (GUM, 2.3.3). In other words:

- the uncertainty evaluation is of type B when the pdf of the error and all the parameters that characterize it are obtained *from information about the measurement procedure and from the measurement value y only*;
- the uncertainty evaluation is of type A when *a statistical analysis of a set of measurements y_n is needed in order to obtain one or more parameters of the pdf.*

Typically, type B evaluations are those relying on the measuring instruments' metrological specifications and are therefore more common whenever they are available and complete. Type A evaluations are common when information about the measuring devices used is missing or insufficient; in such cases, one must autonomously perform a statistical analysis to characterize the measurement error.

In this book, we will make use only of instruments with metrological specifications and therefore of type B evaluations. In order to deal with type A evaluations, it is necessary at least to introduce Student's t-distribution, a topic that falls outside the scope of this book.

1.4.3 Standard Uncertainty

The standard uncertainty is (GUM, 2.3.1) the "uncertainty of the result of a measurement expressed as a standard deviation." The standard uncertainty of the measurement result y is denoted by the lowercase letter u, i.e., $u(y)$. Its definition is specified in greater detail below.

For a *type B evaluation*, the standard uncertainty of the measurement y is the standard deviation of the pdf that characterizes the measurement results, i.e., in our discussion, the standard deviation of the error pdf:

$$u(y) = \text{std}[e] \, . \tag{1.11}$$

The definition of standard uncertainty in a type A evaluation involves Student's t-distribution, so we do not discuss it here.

For the two error pdfs considered above, it is simple to give the values of standard uncertainty:

$$e \sim N(0, \sigma^2) \Rightarrow u(y) = \sigma \, , \tag{1.12}$$

$$e \sim \text{Unif}(-U_{max}, U_{max}) \Rightarrow u(y) = U_{max}/\sqrt{3} \, . \tag{1.13}$$

1.4.4 Expanded Uncertainty, Coverage Probability, Coverage Factor

The expanded uncertainty is a "quantity defining an interval about the result of a measurement that may be expected to encompass a large fraction of the distribution of values that could reasonably be attributed to the measurand" (GUM 2.3.5).

The expanded uncertainty $U(y)$ defines:

- a symmetric interval for the measurement error $(-U(y) \le e \le +U(y))$;
- a probability $p = \text{Prob}(-U(y) \le e \le +U(y))$ associated to the interval, which is called the *coverage probability*[1], and must be conveniently large ("a large fraction of the distribution");
- a ratio $k = U(y)/u(y)$, between the expanded and the standard uncertainties, called the *coverage factor*.

Therefore, given the standard uncertainty, the expanded uncertainty is

$$U(y) = k \cdot u(y) \, . \tag{1.14}$$

The expanded uncertainty is the one used to state the values attributed to the measurand. The GUM specifies that together with the expanded uncertainty, one should always specify the coverage probability p, and also the coverage factor k. "The coverage factor k is always to be stated, so that the standard uncertainty of the measured quantity can be recovered for use in calculating the combined standard uncertainty of other measurement results that may depend on that quantity" (GUM, 3.3.7).

We examine the computation of the expanded uncertainty for the two error pdfs considered above, i.e., the Gaussian pdf and the uniform one.

---

[1]The GUM uses also the term "level of confidence," with the warning that it must not be taken as synonymous with the term "confidence level" used in the theory of confidence intervals. "The terms confidence interval (C.2.27, C.2.28) and confidence level (C.2.29) have specific definitions in statistics and are only applicable to the interval defined by U when certain conditions are met" (GUM, 6.2.2). In order to avoid confusion, we will always use the term "coverage probability."

1.4.4.1 Expanded Uncertainty for Error with Normal (Gaussian) pdf

In this common case, formulas (1.5) and (1.6) in Sect. 1.3.2.1 give the coverage probability p for a given coverage factor k, and conversely. Tables 1.1 and 1.2 give examples of common pairs of coverage probabilities and coverage factors. The typical choice is to compute a 95% expanded uncertainty $U(y)$, by multiplying the standard uncertainty $u(y) = \sigma$ by the coverage factor $k = 1.96 \cong 2$. It is worth noting that in this case, a 100% coverage probability leads to a useless evaluation, $U(y) = +\infty$.

1.4.4.2 Expanded Uncertainty for Error with Uniform pdf

In this case, also quite common, it is possible and useful to consider a 100% coverage probability. The corresponding expanded uncertainty is

$$U(y) = U_{max} . \tag{1.15}$$

The associated coverage factor is $k = U(y)/u(y) = \sqrt{3}$. In this case, of course, the expanded uncertainty is also the maximum absolute value of the measurement error:

$$U = max|e| . \tag{1.16}$$

In contrast to the case of the Gaussian distribution, there is little point in considering an uncertainty with 95% coverage probability: it is negligibly smaller than U_{max}, and leaves a 5% probability of an error outside the specified interval. Therefore, in this case, the choice $U(y) = U_{max}$ is the only sensible one from a practical viewpoint.

1.4.4.3 Accuracy of Instruments, Uniform Distribution of the Error, and "Worst-Case" Uncertainty (WCU)

We have already observed that instrument manufacturers quantify "accuracy" by providing a bound to measurement errors (even if the words "error" and "uncertainty" are usually avoided). In this very common case:

- since no further information is given about the error, it is customary to assume a uniform distribution (as recommended in GUM 4.3.7);
- the provided value of accuracy can be interpreted as a 100% uncertainty (GUM 4.3.7 specifies: "the probability that X_i lies outside this interval is essentially zero").

Therefore, when working with instruments' specifications, it is very common to work with a uniformly distributed error and an uncertainty with coverage probability

$p = 100\%$. We will refer to such an uncertainty with the expression "worst-case uncertainty," or WCU.[2]

1.5 Uncertainty Propagation

1.5.1 Indirect Measurements: Measurement Model, Input Quantities, Output Quantities

The topic of uncertainty propagation is very general with many interesting aspects. Again, a complete discussion goes beyond the scope of this book. We consider here the simple case of a scalar function of N quantities, of the kind

$$\theta = f(x_1, \ldots, x_N) \tag{1.17}$$

In this case a measurement of the quantity θ is obtained by measuring the quantities x_n, $n = 1, \ldots, N$, and is therefore often called "indirect measurement." The function $f(\cdot)$ is called a "mathematical model of the measurement" (GUM, 3.4.1), or "measurement model"; $x_1, \ldots, x_N$ are the "input quantities," and θ is the "output quantity."

We give here an example, taken from the GUM (H.2.2). Consider a circuit element powered by AC sinusoidal voltage with amplitude V, current I, and phase-shift φ. The resistance and the reactance of the elements are given by $R = \frac{V}{I} \cos(\varphi)$, $X = \frac{V}{I} \sin \varphi$. In this case, $N = 3$, and the input quantities are $[x_1; x_2; x_3] = [V; I; \varphi]$. The output quantity is $\theta = R$ and $\theta = X$, for the two indirect measurements, respectively. In many cases it is necessary to consider a set of output quantities, like R and X in this example, as a vector quantity. In the measurements presented in this book, however, this is not required.

The uncertainty of the indirect measurement θ is also called, in the GUM, "combined uncertainty" (GUM 0.7).

1.5.2 Law of Propagation of Errors (LPE)

Of course, we do not have the true values of the input quantities, x_n, but the measured values

$$y_n, \ n = 1, \ldots, N, \tag{1.18}$$

and the unknown associated errors are

---

[2]The expression "worst-case uncertainty" is not used in the GUM, but is more compact than "uncertainty with 100% coverage probability," is clearer than "100% uncertainty," and finally, it is used in the scientific and technical literature [7–10].

$$e_n = y_n - x_n, \quad n = 1, \ldots, N \, . \tag{1.19}$$

We consider as best estimates of the output quantities the values

$$\hat{\theta} = f(y_1, y_2, \ldots, y_n) \, , \tag{1.20}$$

and we seek an expression for the error in the output quantities

$$e_\theta = \hat{\theta} - \theta \, . \tag{1.21}$$

If the nonlinear function $f(\cdot)$ can be approximated with the linear model obtained using the Taylor series up to its first-order term, we can write

$$\hat{\theta} = f(y_1, y_2, \ldots, y_N) =$$
$$= f(x_1 + e_1, \ldots, x_n + e_N) \cong f(x_1, \ldots, x_N) + \sum_{n=1}^{N} c_n e_n \, . \tag{1.22}$$

In this expression, the terms

$$c_n = \frac{\partial \theta}{\partial x_n}, \quad n = 1, \ldots, N \, , \tag{1.23}$$

are called *sensitivity coefficients*. The derivatives are intended to be calculated in the measured values y_n, since the true values x_n are unknown.

On the basis of (1.21) and (1.22), we can write the law of propagation of errors (LPE):

$$e_\theta \cong \sum_{n=1}^{N} c_n e_n \, . \tag{1.24}$$

In all the propagation laws, we write the symbol "$\cong$" because in general we are making use of the first-order Taylor approximation. The relations are exact, and "$\cong$" is substituted by "$=$" when $\theta = f(x_1, \ldots, x_N)$ is a linear relation. It is useful to note that assuming that the input errors e_n have expected value equal to zero, like those considered above, the output error e_θ also has expected value equal to zero as long as the first-order approximation is acceptable.

1.5.3   Law of Propagation of Uncertainty (LPU)

With the expression "law of propagation of uncertainty" the GUM refers to *standard uncertainties*. Standard uncertainties are indeed the ones for which the propagation is easy to calculate in almost every case.

First of all, we write the LPU in the general case, when there are arbitrary correlations between the measurement errors. In this case, the LPU can be written as

$$u^2(\hat{\theta}) = \sum_{n=1}^{N} \sum_{m=1}^{N} c_n c_m u(y_n) u(y_m) \rho(y_n, y_m) =$$
$$= \sum_{n=1}^{N} c_n^2 u^2(y_n) + 2 \sum_{n<m} c_n c_m u(y_n) u(y_m) \rho(y_n, y_m), \quad (1.25)$$

where $u(y_n)$ denotes the standard uncertainty of the measurement y_n, and $\rho(y_n, y_m)$ the correlation coefficients of the errors on y_n and y_m.

The most frequent and simplest form of the LPU is with uncorrelated errors, i.e., $\rho(y_n, y_m) = 0$:

$$u^2(\hat{\theta}) = \sum_{n=1}^{N} c_n^2 u^2(y_n). \quad (1.26)$$

In short, when the errors are uncorrelated, the squared standard uncertainty of the output is a linear combination of the squared standard uncertainties of the input, and the coefficients of the linear combination are the squared sensitivity coefficients.

1.5.4 Propagation of Worst-Case Uncertainties

It is often of practical interest to compute the output WCU $U(\hat{\theta})$, given the input WCUs $U(y_n)$. This computation is feasible only in the case, quite frequent in practice, of *independent input errors*. If (i) the errors e_n are independent, (ii) the input WCUs are $max|e_n| = U(y_n)$, and (iii) the output WCU is $max|e_\theta| = U(\hat{\theta})$, than the propagation of worst-case uncertainties is given by

$$U(\hat{\theta}) = \sum_{n=1}^{N} |c_n| U(y_n). \quad (1.27)$$

WCU propagation consists therefore in directly adding up the WCUs of the input measurements, multiplied by the absolute values of the sensitivity coefficients.

WCU propagation *is not contemplated in the GUM*. Some reasons to avoid computing the WCU are the following:

- the final result makes sense only if all the input errors have a finite WCU;
- the computation is feasible only in the special case of independent errors (it is not sufficient that the errors are uncorrelated);

- most importantly, the result $U(\hat{\theta})$ of WCU propagation, being an uncertainty with 100% coverage probability, may be too pessimistic.

The last statement is a consequence of the well-known central limit theorem of probability theory. According to the LPE expressed by (1.24), indeed, the output error e_θ is the sum of N contributions $c_n e_n$. Therefore, there are essentially two possible cases:

1. If there are many meaningful independent contributions, the pdf of the output error e_θ is bell-shaped, and very similar to a Gaussian distribution. Consequently, $U(\hat{\theta})$ given by (1.27) can be too pessimistic. In this case, a more useful uncertainty evaluation is obtained by propagating the standard uncertainty and then computing the expanded uncertainty $U(\hat{\theta}) = k \cdot u(\hat{\theta})$ with a convenient coverage factor k (typically, $k \cong 2$ for a 95% coverage probability).
2. If, in contrast, there are only one or two meaningful contributions, the pdf of the output error is similar to a rectangle or a trapezoid. In this case, WCU is a simpler and even more correct quantification of the uncertainty.

We illustrate the two cases with two examples. Consider the indirect measurement

$$\theta = \frac{1}{5} \sum_{n=1}^{5} x_n .$$

For the indirectly measured value $\hat{\theta}$, we compute the WCU and the expanded uncertainty with coverage probability 95%, denoting them by $U_{100\%}(\hat{\theta})$ and $U_{95\%}(\hat{\theta})$, respectively. The latter is computed assuming that the distribution of the output error is approximately Gaussian and using therefore the coverage factor $k = 2$.

We suppose input errors with uniform distributions, i.e., $e_n \sim \text{Unif}(-U_n; U_n)$. The sensitivity coefficients are, of course, $c_n = \partial\theta/\partial x_n = 1/5$. As an example of case 1 above, we consider independent identically distributed errors, with $U_n = U_{max} = 1$, $n = 1, \ldots, 5$. By easy computations, considering that $U(y_n) = U_n, u(y_n) = U_n/\sqrt{3}$, we have

$$U_{100\%}(\hat{\theta}) = \sum_{n=1}^{N} |c_n| U(y_n) = 1 ,$$

$$u(\hat{\theta}) = \sqrt{\sum_{n=1}^{N} c_n^2 u^2(y_n)} = 0.2582 ,$$

$$U_{95\%}(\hat{\theta}) = 2 \cdot u = 0.5164 .$$

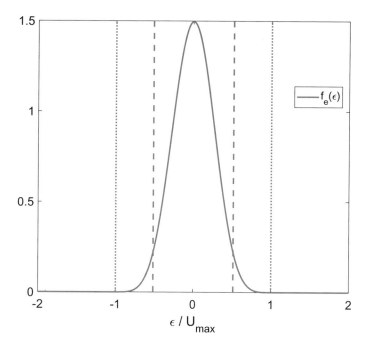

Fig. 1.5 Pdf of the output error in case 1. Dotted lines indicate the WCU, dashed lines the expanded uncertainty $U = k \cdot u$, with $k = 2$. The WCU is clearly pessimistic

In this case, the WCU appears clearly too pessimistic with respect to the uncertainty with 95% probability. The distribution of the output error, with the computed uncertainties, is shown in Fig. 1.5.

As an example of case 2. we consider the very same problem with $U_1 = 4$, $U_2 = 0.7$, $U_3 = U_4 = U_5 = 0.1$. In this case, we have

$$U_{100\%}(\hat{\theta}) = \sum_{n=1}^{N} |c_n| U(y_n) = 1 \,,$$

$$u(\hat{\theta}) = \sqrt{\sum_{n=1}^{N} c_n^2 u^2(y_n)} = 0.4693 \,,$$

$$U_{95\%}(\hat{\theta}) = 2 \cdot u = 0.9387 \,.$$

In this case, the WCU is not pessimistic. Instead, $U_{95\%}$ is pessimistic, in the sense that it is very close to the WCU, and the exact associated coverage probability is clearly higher than 95%. The pdf of the output error for this case is represented in Fig. 1.6.

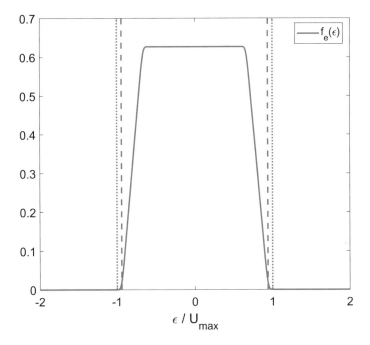

Fig. 1.6 Pdf of the output error in case 2. Dotted lines indicate the WCU $U_{100\%}$, dashed lines the expanded uncertainty $U_{95\%}$. The WCU is adequate, while the probability 95% associated to $U_{95\%}$ is pessimistic (the actual probability is higher and near 100%)

In this book, the propagation of the WCU will be taken into due consideration and used. One main reason is that, as we will see, it is very often used in instruments' specifications. Instrument users indeed prefer that the "accuracy" be a defined number, without a 95% probability attached to it; and instrument manufacturers want to satisfy their customers. In instrument specifications we always read the "accuracy" as a direct sum of independent contributions, and never as a root square of a sum of squared contributions multiplied by a coverage factor.

1.5.5 Laws of Propagation in Terms of Relative Quantities

It is often convenient to consider relative errors and uncertainties. According to the GUM (B.2.20), the relative error is the error divided by the true value:

$$e_r = \frac{e}{x}. \tag{1.28}$$

The relative uncertainty is the uncertainty divided by the absolute value of the true error, but in practice, it is evaluated by dividing the uncertainty by the absolute value of the measurement result:

$$U_r = \frac{U}{|x|} \cong \frac{U}{|y|} \,. \tag{1.29}$$

The former definition applies to all kinds of uncertainties: standard, expanded, and worst-case. While errors and uncertainties have the same physical dimension of the measurement (for example volts, amperes), relative quantities are nondimensional and are often expressed as a percentage (%) or in parts per million (ppm), etc.

It is easily demonstrated that the laws of propagation for relative quantities are identical to those for absolute quantities, by substituting the sensitivity coefficients with the *relative sensitivity coefficients*:

$$c_{r,n} = \frac{\partial\theta/\partial x_n}{\theta/x_n} = c_n \cdot \frac{x_n}{\theta} \,. \tag{1.30}$$

It is also immediate to verify that the relative sensitivity coefficients can be computed by taking the derivative of the natural logarithm of the output quantity:

$$c_{r,n} = \frac{\partial \ln(\theta)}{\partial x_n} \cdot x_n \,. \tag{1.31}$$

The propagation laws, in terms of relative quantities, are the following.

LPE:

$$e_{r,\theta} \cong \sum_{n=1}^{N} c_{r,n} e_{r,n} \,. \tag{1.32}$$

LPU (standard uncertainties, uncorrelated errors):

$$u_r^2(\hat{\theta}) = \sum_{n=1}^{N} c_{r,n}^2 u_r^2(y_n) \,. \tag{1.33}$$

The law for correlated errors is analogous to (1.25), substituting relative uncertainties and sensitivity coefficients for absolute ones. The correlation coefficients do not change.

LPU (worst-case uncertainties, independent errors):

$$U_r(\hat{\theta}) = \sum_{n=1}^{N} |c_{r,n}| U_r(y_n) \,. \tag{1.34}$$

1.5.6 Propagation for Some Typical Indirect Measurements

Working with absolute quantities (errors, uncertainties, sensitivity coefficients) is especially easy and convenient in dealing with indirect measurements that are linear combinations of the input quantities, with a possible constant additive term. The measurement model is of the form

$$\theta = k + \sum_{n=1}^{N} a_n x_n . \tag{1.35}$$

In this case, of course, the sensitivity coefficients c_n are the coefficients a_n of the linear combination. Therefore, for this kind of indirect measurements, the propagation formulae are, quite trivially, as follows:

LPE:

$$e_\theta = \sum_{n=1}^{N} a_n e_n . \tag{1.36}$$

LPU (standard uncertainty, uncorrelated errors):

$$u(\hat{\theta}) = \sqrt{\sum_{n=1}^{N} a_n^2 u^2(y_n)} . \tag{1.37}$$

LPU (worst-case uncertainty, independent errors):

$$U(\hat{\theta}) = \sum_{n=1}^{N} |a_n| U(y_n) . \tag{1.38}$$

Working with relative quantities is especially convenient, however, when one is dealing with indirect measurements that are *monomial functions* of the input quantities, that is, of the form

$$\theta = k \prod_{n=1}^{N} x_n^{b_n} . \tag{1.39}$$

In this case, it is readily verified that the relative sensitivity coefficients $c_{r,n}$ are the exponents b_n of the monomial. Consequently, the propagation formulas for a monomial function of the kind (1.39) are the same as those for a linear combination, but with relative quantities instead of absolute ones:

LPE:

$$e_{r,\theta} = \sum_{n=1}^{N} b_n e_{r,n} \, .$$

(1.40)

LPU (standard uncertainty, uncorrelated errors):

$$u_r(\hat{\theta}) = \sqrt{\sum_{n=1}^{N} b_n^2 u_r^2(y_n)} \, .$$

(1.41)

LPU (worst-case uncertainty, independent errors):

$$U_r(\hat{\theta}) = \sum_{n=1}^{N} |b_n| U_r(y_n) \, .$$

(1.42)

Important particular cases of the above formulas are the difference and the ratio of measurements. For a difference of measurements,

$$\theta = x_1 - x_2$$

(1.43)

the propagation formulas are the following:

LPE:

$$e_\theta = e_1 - e_2 \, .$$

(1.44)

LPU (standard uncertainty, uncorrelated errors):

$$u(\hat{\theta}) = \sqrt{u^2(y_1) + u^2(y_2)} \, .$$

(1.45)

LPU (worst-case uncertainty, independent errors):

$$U(\theta) = U(y_1) + U(y_2) \, .$$

(1.46)

For a ratio of measurements

$$\theta = x_1/x_2 \, ,$$

(1.47)

the propagation formulas are the following.

LPE:

$$e_{r,\theta} = e_{r,1} - e_{r,2} \, .$$

(1.48)

LPU (standard uncertainty, uncorrelated errors):

$$u_r(\hat{\theta}) = \sqrt{u_r^2(y_1) + u_r^2(y_2)}\,. \qquad (1.49)$$

LPU (worst-case uncertainty, independent errors):

$$U_r(\theta) = U_r(y_1) + U_r(y_2)\,. \qquad (1.50)$$

The formulas for the ratio are, of course, identical to those for the difference, substituting relative quantities for absolute quantities, because the relative sensitivity coefficients $c_{r,n}$ for the ratio are equal to the absolute ones for the difference (1 and -1).

1.5.7 A Note About Independence of Errors

The hypothesis of independent errors may appear quite restrictive, and indeed, it is not met in some practical cases, e.g., signals with additive colored noise. However, it is very common in practice when one is working with instruments. A typical example is shown in Fig. 1.7. If the input quantities are measured using different instruments, then measurement errors must be considered independent. Therefore, both the formulas for worst-case uncertainty propagation and those for standard uncertainty propagation of uncorrelated errors apply.

Indeed, in the case of input quantities measured by the very same instruments, under due hypotheses (essentially, measurements obtained with the same instrument

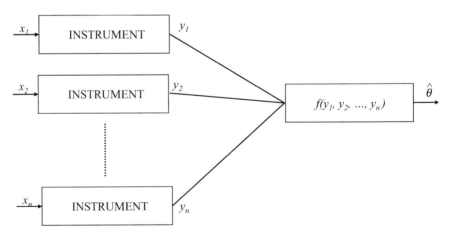

Fig. 1.7 A typical case of propagation with independent errors: input quantities measured using different instruments

settings and in a short time interval), the introduced errors may also be decomposed into independent components. This fact is examined in the following section.

1.6 Errors and Uncertainty in Instruments

1.6.1 Measuring Instrument as a Black Box

Every measuring instrument can be seen, from the viewpoint of measurement errors, as a black box, and it can be represented as in Fig. 1.2, substituting the word "instrument" for the GUM expression "measurement procedure" (see Fig. 1.8).

For an ideal instrument, the measurement error is zero, and the input–output characteristic is represented by the straight line $y = x$ (Fig. 1.9).

The error introduced by a real-world instrument is actually the sum of many different contributions. We discuss these contributions one at a time, briefly illustrating the meaning, and the effect, of each of them.

1.6.2 Ideal Uniform Roundoff Quantization. Quantization Error

The possible output values of a measuring instrument always form a finite set. This is true not only for digital instruments (the ones used in this book), but also for analog instruments. Indeed, any measurement, even if obtained by reading an analog scale, is always represented with a finite (and usually fixed in advance) number of digits.

Fig. 1.8 Measuring instrument seen as a black box

Fig. 1.9 Input–output characteristic of an ideal measuring instrument

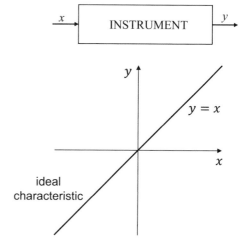

Fig. 1.10 Ideal uniform
roundoff quantization: when
the analog value x is within
the interval delimited by the
ideal threshold levels
th_{k-1}^{id}, th_k^{id}, the output
assumes the quantized value
y_k^q

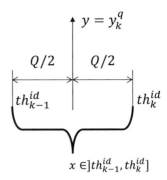

Quantization is therefore inherent in the process of measuring, and the output of an instrument is always quantized.

Let us consider an ideal measuring instrument with quantization. Generally speaking, quantization consists in approximating analog values x with digital (quantized) values y_k^q, $k = 1, \ldots, M$, where M is the number of output values. We consider here uniform and roundoff quantization, which is ubiquitous in general-purpose digital instruments. In uniform quantization there are M equispaced output values y_k^q, $k = 1, \ldots, M$, with quantization step

$$Q = y_{k+1}^q - y_k^q, k = 1, \ldots, M - 1. \tag{1.51}$$

The quantization step is also called the "ideal code bin width" in the IEEE Standard [11], which is specific for digitizing waveform recorders. It is useful to highlight that the fundamental concepts of this standard are general, powerful, and applicable also to different kinds of instruments. Roundoff quantization associates the kth quantized output y_k^q to any analog value within the interval delimited by the "ideal" threshold levels $th_{k+1}^{id} = y_k^q - Q/2, th_k^{id} = y_k^q + Q/2$, as illustrated in Fig. 1.10. The threshold levels are also called "code bin transition levels" in the IEEE Standard [11].

We call "ideal" both the levels $th_k^{id}, k = 1, \ldots, M - 1$, and the associated uniform roundoff quantization: this is useful in that it stresses the difference with respect to what happens in real-world instruments, where the thresholds are not exactly in their ideal positions (as discussed in the following section).

In order to define fully the quantization characteristic, it must be specified that the first output level, y_1^q, is associated to an input $x < th_1^{id}$, and the last level, y_M^q, to an input $x > th_{M-1}^{id}$. Moreover, it is also useful to define the values $x_{min} = y_1^q - Q/2$, $x_{max} = y_M^q + Q/2$, which delimit the full-scale range of the quantizer. The sequence of output levels and threshold levels, with the additional quantities x_{min}, x_{max}, is illustrated in Fig. 1.11.

We indicate the operation of ideal uniform roundoff quantization with the function

$$y = quant(x).$$

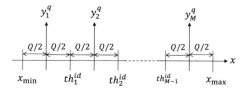

Fig. 1.11 Ideal uniform roundoff quantization: sequence of output quantized values y_k^q and (ideal) threshold levels th_k^{id}. The first quantized output level is y_1^q, and the level $x_{min} = y_1^q - Q/2$ marks the beginning of the input range of the quantizer

Fig. 1.12 Input–output characteristic of an ideal uniform roundoff quantizer and associated quantization error. In the figure, the number of quantized output levels is $M = 8$

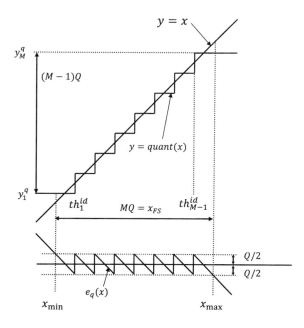

This function, or input–output characteristic, is represented in Fig. 1.12, together with the associated quantization error

$$e_q(x) = quant(x) - x .$$

It is worth noting that the ideal threshold levels define $M - 2$ equal quantization steps along the x-axis:

$$Q = th_{k+1}^{id} - th_k^{id}, k = 1, \ldots, M - 2, \tag{1.52}$$

just as the output levels y_k^q define $M - 1$ equal quantization steps along the y-axis, as expressed by (1.51).

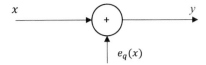

Fig. 1.13 Model of a measuring instrument affected by quantization error only. Quantization error can be seen as a spectrally white (constant power density spectrum) additive noise with uniform distribution.

The characteristic $y = quant(x)$ is a perfectly regular staircase, whose steps have width and height equal to Q. It approximates, as closely as possible, the ideal characteristic without quantization, $y = x$. The quantization error $e_q(x)$ has maximum absolute value, or worst-case quantization uncertainty,

$$U_q = max|e_q(x)| = Q/2,$$

as long as the input x is within the interval $[x_{min}, x_{max}]$. The full-scale range of the quantizer is $x_{FS} = x_{max} - x_{min} = MQ$.

For example, in an 8-bit digital oscilloscope with a full-scale range $x_{FS}=8$ V (corresponding to 1 V/div with eight vertical divisions), we have

$$M = 2^8 = 256,$$

$$Q = x_{FS}/M = 8/256 = 0.03125 \text{ V},$$

$$U_q = Q/2 = 0.015625 \text{ V}.$$

Quantization error can be seen as an additive "noise" added to the input signal. The noise, under broad assumptions, is uniformly distributed, i.e., $e_q \sim \text{Unif}(-U_q, U_q)$, and is spectrally white [12]. Therefore, one can create a simple model of an instrument affected by quantization error only, like the one shown in Fig. 1.13.

Since the quantization error is uniformly distributed, the standard quantization uncertainty is

$$u_q = \frac{U_q}{\sqrt{3}} = \frac{Q}{\sqrt{12}}.$$

1.6.3 Nonideal Uniform Roundoff Quantization

A real-world instrument differs, of course, from an ideal quantizer. The first and most important difference is that the actual threshold levels, th_k, are different from the ideal one th_k^{id}. We call

$$y = nlquant(x)$$

Fig. 1.14 Model of the nonideal quantization *nlquant* (x) occurring in actual measuring instruments. The nonlinear transformation $g(x)$ transforms the ideal thresholds th_k^{id} into the actual one th_k

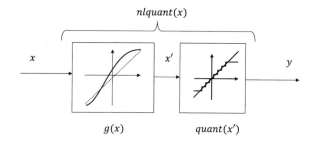

the transformation associated with the quantization with threshold levels $th_k \neq th_k^{id}$. The actual thresholds define the actual x-axis quantization steps

$$Q_k = th_{k+1} - th_k, k = 1, \ldots, M - 2. \qquad (1.53)$$

This equation corresponds to (1.52) for ideal quantization, which gives the ideal quantization step Q. The quantities Q_k are also called (actual) "code bin widths" in the IEEE Standard [11].

The function *nlquant* (x) can be mathematically modeled as a continuous nonlinear transformation $g(x)$ cascaded with the ideal quantization, as shown in Fig. 1.14. If th_k^{id} are the ideal thresholds and th_k the actual thresholds, then the function $g(x)$ is uniquely defined only at the points $x = th_k$, by the relation

$$g(th_k) = th_k^{id}, k = 1, \ldots, M - 1. \qquad (1.54)$$

The other requirement on $g(x)$ is that for x in the interval $]th_{k-1}, th_k]$, $g(x)$ must be in the interval $]th_{k-1}^{id}, th_k^{id}]$. If this requirement is met, together with (1.54), we have the input–output transformation $y = nlquant(x)$, that is,

$$y = y_k^q \Leftrightarrow x' = g(x) \in]th_{k-1}^i d, th_k^{id}] \Leftrightarrow x \in]th_{k-1}, th_k].$$

An example of the transformation $y = nlquant(x)$, with two possible associated functions $g(x)$, is represented in Fig. 1.15. One of the functions $g(x)$ is piecewise linear, which is, in a sense, the most natural choice.

The illustrated model takes account of the error introduced by a real-world measuring instrument under conditions of negligible noise, i.e., of the systematic part of the error. It does not model some sources of systematic error (for example, hysteresis in the input–output characteristic), but is general enough for most general-purpose laboratory instruments. Now we examine how the model allows a natural, and widely used, decomposition of the error into different contributions, corresponding to gain, offset, and nonlinearity errors.

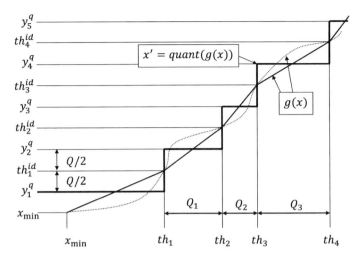

Fig. 1.15 Non-ideal quantization $nlquant(x)$ resulting from the cascade of $x' = g(x)$ and $y = quant(x')$. The result is a staircase with step widths $Q_k \neq Q$. Two possible functions $g(x)$ producing the same effect are represented (one is piecewise linear)

1.6.4 Gain, Offset, Integral Nonlinearity Errors

It is useful to express the nonlinear transformation $g(x)$ in the form

$$g(x) = Gx + O + inl(x),$$

where G is the gain, O is the offset, and $inl(x)$ the integral nonlinearity.

The meaning of this expression is illustrated in Fig. 1.16. The straight line $y = Gx + O$, which we call the *reference line*, approximates the curve $g(x)$ and is chosen to minimize, in some sense, the integral nonlinearity $inl(x)$. The reference line may be chosen freely, according to the particular circumstances. It can even coincide with the ideal characteristic or can have the offset term equal to zero or can minimize the maximum nonlinearity $max|inl(x)|$, etc.

The difference between the gain G and its ideal value is the (relative) gain error. If we consider devices such as sensors and amplifiers, the ideal value of G may be different from the unit value. Since we are considering measuring instruments, in which the ideal characteristic is $y = x$, the ideal gain is $G = 1$, and the gain error is

$$e_{r,G} = G - 1 = \Delta G.$$

The gain error is relative because it is nondimensional, and coincides with the *relative* error of the instrument, if the only nonideal behavior is $G \neq 1$.

The difference between the offset O and its ideal value is the (absolute) offset error. In the case of measuring instruments, the ideal value is zero, and therefore

Fig. 1.16 Decomposition of $g(x)$ into the sum of a reference line $y = Gx + O$ and an integral nonlinearity $inl(x)$

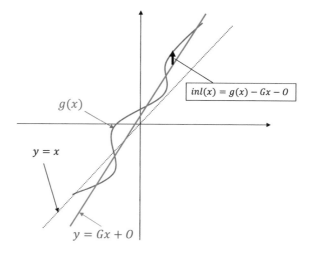

$$e_O = O - 0 = O.$$

We can now express the error introduced by the transformation $g(x)$. It is the sum of the gain error, offset error, and integral nonlinearity error:

$$e(x) = g(x) - x = e_{r,G} \cdot x + e_O + inl(x).$$

The term $e_{r,G} \cdot x = e_G(x)$ is the *absolute* gain error.

It is useful to note that just as $g(x)$ is uniquely defined only at the abscissas th_k, the integral nonlinearity is also defined uniquely at the same abscissas. The uniquely defined values of integral nonlinearity are

$$inl_k = inl(th_k), k = 1, \ldots, M - 1.$$

Another way to write the integral nonlinearity is

$$inl_k = [g(x) - Gx - O]_{x=th_k} = th_k^{id} - G \cdot th_k - O.$$

It is therefore the difference between the ideal and the actual threshold levels, after correcting for gain and offset of the instrument. To assess the quality of a nonideal quantizer (usually an analog-to-digital converter), one also uses the *differential non-linearity*, which is the discrete derivative of inl_k:

$$dnl_k = inl_{k+1} - inl_k, k = 1, \ldots, M - 2.$$

This quantity corresponds to the error in the width of the kth x-axis quantization step, after correcting for the gain of the instrument:

$$dnl_k = (th_{k+1}^{id} - th_k^{id}) - G \cdot (th_{k+1} - th_k) = Q - G \cdot Q_k.$$

Conversely, inl_k is the discrete integral of dnl_k, whence the name "integral nonlinearity."

1.7 Error Model and Uncertainty Specifications for a Measuring Instrument

1.7.1 Error Model

The error components discussed above can be represented in an additive error model of the measuring instrument, sketched in Fig. 1.17.

The error introduced by the instrument is therefore

$$e(x) = e_{r,G} \cdot x + e_O + inl(x) + e_q(x'). \tag{1.55}$$

In order to understand the meaning of each error component, it is useful to consider their qualitative effect on an input signal $x(t)$.

Gain error $e_{r,G}$ introduces an unwanted amplification or attenuation of the signal (depending on whether $e_{r,G} > 0$ or < 0). It does not alter the shape of the signal and does not affect *ratios* between its values. For example, the ratio $max(x(t))/min(x(t))$ is not changed by the presence of a gain error.

Offset error e_O introduces an unwanted positive or negative shift of the signal (depending on whether $e_O > 0$ or < 0). This error, too, does not alter the shape of the signal. Moreover, it does not affect *differences* between its values; for example,

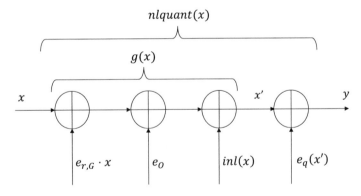

Fig. 1.17 Additive error model for a measuring instrument, including gain, offset, integral nonlinearity, and quantization error. It must be noted that $e_G = e_{r,G} \cdot x$ and inl are functions of the instrument input x, while e_q is a function of the transformed input x'

the difference $max(x(t)) - min(x(t))$ is not changed by the presence of an offset error.

Integral nonlinearity $inl(x)$, in a sense, does not amplify or attenuate the signal, nor does it introduce a positive or negative shift, because these effects are taken into account by gain and offset errors. Instead, it *alters the shape of the signal*, i.e., it introduces distortion. If the signal is a sinusoid, the integral nonlinearity, unlike gain and offset errors, introduces spurious harmonics. The ratio and differences between values of the signal are altered by the integral nonlinearity.

Quantization error $e_q(x')$ is very similar to integral nonlinearity, since it is actually a particular kind of nonlinearity. The difference, with respect to $inl(x)$, is essentially that its magnitude is determined only by the quantization step Q. It can be added that $inl(x)$ is mainly a "large-scale" nonlinearity (which can be approximated by a smooth curve), while $e_q(x')$ is inherently a "small-scale," or "granular," nonlinearity (consisting of a high number of small discontinuities, which cannot be approximated by a smooth curve). Therefore, $e_q(x')$ adds to a sinusoidal signal $x(t)$ a very large number of small harmonics, while $inl(x)$ mainly adds a smaller number of higher harmonics. Of course, $e_q(x')$, just like $inl(x)$, alters ratios and differences of values of the signal.

The errors discussed above are of a systematic nature: they are uniquely determined by the value x of the measurand. In order to consider also random errors, we can introduce an equivalent noise source at the input of the instrument. A reasonably complete error model,[3] for a general measuring instrument, is therefore that in Fig. 1.18. The equivalent input noise is denoted here with a capital letter, E_{noise}, to highlight its random nature. The total error is

$$e_{tot} = E_{noise} + e(x). \qquad (1.56)$$

A very important feature of the illustrated error model is that the error sources $e_{r,G}, e_O, e_{inl}, e_q, E_{noise}$ can be considered *independent*. For this reason, we are able:

- to compute worst-case uncertainties (if the noise contribution is negligible or is characterized by a WCU), using the propagation formula (1.27);
- to compute standard uncertainties, using the simple propagation formula (1.26), which does not involve correlation coefficients.

1.7.2 Uncertainty Specifications

The specifications for gain, offset, and nonlinearity errors are usually given in terms of worst-case uncertainties, i.e.,

---

[3]The illustrated model does not take into account *dynamic* errors, due, for example, to the limited bandwidth of the instrument. They can be modeled by a linear system (typically a lowpass filter) at the instrument input. If the bandwidth of the input signal $x(t)$ is conveniently lower than the instrument bandwidth, dynamic errors can be neglected.

$$max|e_{r,G}| = U_{r,G} \,,$$

$$max|e_O| = U_O \,,$$

$$max|inl| = U_{inl} \,.$$

It is useful to remember that according to the GUM, when the only information available about the error is the maximum absolute value U_{max}, one must attribute to it a uniform distribution, $e \sim \text{Unif}(-U_{max}, U_{max})$. Since the error distributions are uniform, the standard uncertainties are $u_{r,G} = U_{r,G}/\sqrt{3}$, $u_O = U_O/\sqrt{3}$, $u_{inl} = U_{inl}/\sqrt{3}$. To these specifications must be added the one concerning quantization:

$$max|e_q| = U_q = Q/2 \,,$$

with the standard uncertainty $u_q = U_q/\sqrt{3}$.

Finally, the equivalent input noise is usually specified in terms of root mean square value, i.e., the standard deviation

$$\text{std}[E_{noise}] = \sigma_{noise} = u_{noise} \,.$$

When the only information available about the error is its standard deviation σ, one must attribute to it a normal distribution, $e \sim N(0, \sigma^2)$. This is also in accordance with the typical frequentist distribution of the noise.

Before examining practical examples of such specifications, it is useful to consider the expression of the uncertainty of typical measurements in terms of the specifications listed above.

1.7.3 Uncertainty of Typical Measurements

We derive formulas of worst-case and standard uncertainties for typical measurements. WCUs, as discussed in Sects. 1.4.4.3 and 1.5.4, make sense only when the noise contribution in the model of Fig. 1.18 is negligible (u_{noise} is identically zero or has a finite maximum value). If a meaningful Gaussian contribution is present,

Fig. 1.18 Additive error model for a measuring instrument, including both static systematic errors and random errors. The latter are represented by the equivalent input noise source E_{noise}

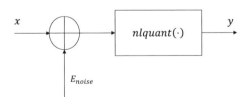

WCU is infinite, and one must compute instead the standard uncertainty (obtaining from it the expanded uncertainty, by applying a convenient coverage factor).

It is necessary to clarify that in order to follow the GUM, one should not use worst-case uncertainties at all. There is, however, a very compelling reason to compute them: understanding the specifications of real-world instruments. As already mentioned, such specifications are always given as sums of worst-case components, as in propagation formula (1.27), and never as quadratic sums of standard deviations, as in propagation formula (1.26).

When computing WCUs, we will hypothesize that the noise is negligible. When computing standard uncertainties, we can discard this hypothesis, and we will include u_{noise} in the result.

1.7.3.1 Uncertainty of a Direct Measurement

From the error model (1.55)–(1.56), and considering that the error contributions are independent, the WCU $U(y) = max|e|$ is readily computed:

$$U(y) = U_{r,G} \cdot |y| + U_O + U_{inl} + U_q . \tag{1.57}$$

The relative WCU is

$$U_r(y) = U_{r,G} + \frac{U_O + U_{inl} + U_q}{|y|} .$$

The expression of the standard uncertainty is equally simple, and it can include the contribution u_{noise} if it is not negligible:

$$u(y) = \sqrt{u_{r,G}^2 \cdot y^2 + u_O^2 + u_{inl}^2 + u_q^2 + u_{noise}^2} .$$

The relative standard uncertainty is

$$u_r(y) = \sqrt{u_{r,G}^2 + \frac{u_O^2 + u_{inl}^2 + u_q^2 + u_{noise}^2}{y^2}} .$$

1.7.3.2 Uncertainty of a Difference of Measurements

For the quantity $\theta = x_1 - x_2$, the error, according to LPE expressed by (1.36), is $e_\theta = e(x_1) - e(x_2)$. Let us compute, first, the WCU, supposing that the errors are independent, and the uncertainties $U_{r,G}$, U_O, U_inl, U_q are the same for the two measurements. In this case, we have

$$U(\hat{\theta}) = U(y_1) + U(y_2) = U_{r,G} \cdot (|y_1| + |y_2|) + 2(U_O + U_{inl} + U_q) . \tag{1.58}$$

This formula, however, is not correct, since the errors $e(x_1)$ and $e(x_2)$ are not independent. If x_1 and x_2 are measured:

(1) by the same instrument;
(2) with the same settings;
(3) in a short time interval;

we can assume that they are measured with the same input–output characteristic $nlquant(\cdot)$, and in particular with the same gain error $e_{r,G}$ and the same offset error e_O.

Under this hypothesis, we first write the expression of the error $e_\theta = e(x_1) - e(x_2)$, including the contributions $E_{noise}(x_1)$ and $E_{noise}(x_2)$, i.e., the noise errors affecting the measurements of x_1 and x_2, respectively. The result is

$$
\begin{aligned}
e_\theta &= e_{r,G} \cdot x_1 + e_O + inl(x_1) + e_q(x_1') + E_{noise}(x_1) + \\
&\quad - [e_{r,G} \cdot x_2 + e_O + inl(x_2) + e_q(x_2') + E_{noise}(x_2)] = \\
&= e_{r,G} \cdot (x_1 - x_2) + [inl(x_1) + e_q(x_1') + E_{noise}(x_1)] + \\
&\quad - \left[inl(x_2) + e_q(x_2') + E_{noise}(x_2) \right] .
\end{aligned}
\tag{1.59}
$$

Now we can write the WCU (supposing $E_{noise} \cong 0$):

$$
U(\hat\theta) = max|e_\theta| = U_{r,G}|y_1 - y_2| + 2(U_{inl} + U_q) .
\tag{1.60}
$$

In this expression, quantization errors affecting the two measurements have been considered independent, which is the right thing to do. Even if $y_1 = y_2$, we cannot say anything about the quantization errors $e_q(x_1')$ and $e_q(x_2')$, apart from the fact that their maximum value is U_q. The same applies to $inl(x)$ if no other information is available about this function.

By comparing the derived expression with (1.58), we note that offset error, as expected, does not contribute to the uncertainty. Furthermore, gain uncertainty multiplies $|y_1 - y_2|$, and not $|y_1| + |y_2|$. Summing up, knowing the separate independent uncertainty contributions (gain, offset, integral nonlinearity, quantization, noise) leads to a lower, and correct, uncertainty evaluation for difference measurements.

It must be noted in particular that the uncertainty (1.60) is lower than (1.58), since the contribution of U_O is missing, and it is much lower if $|y_1 - y_2| \ll |y_1| + |y_2|$. This situation occurs when one is measuring a small difference of large values.

From formula (1.59) is also readily computed the standard uncertainty:

$$
u(\hat\theta) = \sqrt{u_{r,G}^2 \cdot (y_1 - y_2)^2 + 2 \cdot (u_{inl}^2 + u_q^2 + u_{noise}^2)} .
$$

Also in $u(\hat\theta)$, as in (1.60), the contribution of the offset uncertainty is zero, and the gain uncertainty multiplies the square of the difference of the measurements (and not the sum of the squares, as for independent measurement errors).

It is possible to derive formulas similar to those for differences of measurements for any other indirect measurement, e.g., for ratios (quantities of the form $\theta = x_1/x_2$), ratios of differences ($\theta = (x_1 - x_2)/(x_3 - x_4)$). Deriving and illustrating such formulas is beyond the scope of our discussion and is left to the interested reader.

1.8 Examples of Manufacturer's Uncertainty Specifications of Laboratory Instruments

Below, we give examples of uncertainty specifications of real commercial instruments and their interpretation in terms of gain, offset, integral nonlinearity, and quantization uncertainties. It is necessary to specify that the figures extracted by the specifications, although logical and derived by careful considerations from the instruments' data sheets, are necessarily partial. They are not meant to quantify accurately and in full the performance of the considered instruments; such a task would require taking into account the entire data sheet. The interpretations provided are useful, rather, for understanding the theory presented here in connection with real-world instruments, and therefore also for reading the data sheets with a better appreciation of the instruments' performance.

1.8.1 Example of Uncertainty Specifications for the Tektronix TDS1000B–TDS2000B Series Oscilloscopes

1.8.1.1 Vertical (Voltage) Measurements

The oscilloscopes of this series have a vertical resolution of eight bits ($M = 256$), and the vertical scale has eight divisions. Therefore, the quantization uncertainty, as a percentage of the full-scale range, is

$$U_q = \frac{Q}{2} = \frac{x_{FS}}{2M} = \frac{x_{FS}}{512} \cong 0.2\% \, x_{FS} \, .$$

The same uncertainty, expressed in number of vertical divisions, is

$$U_q = \frac{x_{FS}}{512} = \frac{8 \text{ div}}{512} \cong 0.0156 \text{ div} \, .$$

Now let us examine an excerpt from the instrument specifications [13] for voltage measurements of interest to us (Fig. 1.19).

Before going into detail, we highlight some facts.

Oscilloscope Specifications (Cont.)

Vertical		
DC Gain Accuracy	±3% for Sample or Average acquisition mode, 5 V/div to 10 mV/div	
	±4% for Sample or Average acquisition mode, 5 mV/div and 2 mV/div	
DC Measurement Accuracy, Average Acquisition Mode	*Measurement Type*	*Accuracy*
	Average of ≥ 16 wave-forms with vertical position at zero	±(3% × reading + 0.1 div + 1 mV) when 10 mV/div or greater is selected (b)
	Average of ≥ 16 wave-forms with vertical position not at zero	±[3% × (reading + vertical position) + 1% of vertical position + 0.2 div] (c) Add 2 mV for settings from 2 mV/div to 200 mV/div Add 50 mV for settings from > 200 mV/div to 5 V/div
Volts Measurement Repeatability, Average Acquisition Mode	Delta volts between any two averages of ≥ 16 waveforms acquired under same setup and ambient conditions	±(3% × reading + 0.05 div) (a)

Fig. 1.19 Excerpt from vertical specifications for TDS1000B–TDS2000B Series oscilloscopes [13]. Parts of interest are highlighted, and the specifications considered are marked (**a**), (**b**), (**c**)

1. The word "uncertainty" does not appear. The word "accuracy" is used. This is a common and understandable practice in describing the metrological performance of commercial instruments.
2. Formulas to compute the "accuracy" are given as direct sums of positive contributions. Therefore, worst-case sums of error contributions are considered.
3. Formulas are given for averages of ≥ 16 waveforms. Averaging makes the small noise present in the instrument negligible. This is consistent with considering worst-case sums of uncertainties.
4. Formulas are given for difference of measurements or "delta volts" (a), and for direct measurements (b), (c).

First, we consider specification (a), which regards differences of measurements. We rewrite formula (1.60) for $\hat{\theta} = y_1 - y_2$:

$$U(\hat{\theta}) = U_{r,G}|y_1 - y_2| + 2\left(U_{inl} + U_q\right). \tag{1.61}$$

By comparing it with (a), we have

$$U_{r,G} = 3\%,$$

$$2(U_{inl} + U_q) = 0.05 \text{ div}.$$

By approximating $U_q \cong 0.015$ div, we have therefore

$$U_{inl} \approx \frac{0.05}{2} - 0.015 = 0.01 \text{ div} .$$

Second, we consider specification (b), which regards direct measurements with vertical position setting $V_P = 0$ and vertical sensitivity $\geq 10\,\text{mV/div}$. We compare it with formula (1.57), rewritten here for convenience:

$$U(y) = U_{r,G} \cdot |y| + U_O + U_{inl} + U_q .$$

The comparison gives, consistently with specification (a), $U_{r,G} = 3\%$. Besides, we have

$$U_O + U_{inl} + U_q = 0.1 \text{ div} + 1 \text{ mV} .$$

Since we have already obtained, from specification (a), that $U_q + U_{inl} \approx 0.025$ div, we have

$$U_O \approx 0.075 \text{ div} + 1 \text{ mV} .$$

The offset uncertainty is the sum of a "fixed" term (1 mV) and a term proportional to the selected range (0.075 div), obviously for technical reasons.

Third, we consider specification (c), which regards direct measurements with vertical position setting $V_P \neq 0$. Setting $V_P \neq 0$ is equivalent to the following sequence of operations:

- measuring the voltage $x + \hat{V}_P$, where $\hat{V}_P$ is the actual analog voltage added, which is different from the nominal vertical position V_P;
- obtaining the intermediate result $y' = nlquant(x + \hat{V}_P)$;
- deriving the final measurement with the operation $y = y' - V_P$.

Consequently, in this case we must expect the gain uncertainty to multiply the sum $y + V_P$, and not y. Moreover, we must expect a higher offset uncertainty, due to the difference between V_P and $\hat{V}_P$ (which is an additional offset error).

Without going too deeply into a mathematical discussion, we see that specification (c) is consistent with our expectations. The gain uncertainty $U_{r,G} = 3\%$ multiplies the sum (reading + vertical position), and the offset uncertainty is higher.

It is important to note that in voltage measurements with digital oscilloscopes, we have a term $U_{r,G}|y|$ for vertical position $V_P = 0$, and a term $U_{r,G} \cdot |y + V_P|$ for $V_P \neq 0$. For this reason, other manufacturers prefer to include the gain uncertainty in the form $U_{r,G} \cdot x_{FS}$, a term that is easier to evaluate and is always higher than the maximum error $U_{r,G} \cdot |y + V_P|$. Summing up, for the examined measurements, we have

$U_{r,G} = 3\%$,
$U_O \approx 0.075 \text{ div} + 1 \text{ mV}$ for $V_P = 0$,
$U_O \approx 0.175 \text{ div} + 1\%|V_P|$ for $V_P \neq 0$,
$U_{inl} \approx 0.01 \text{ div}$,
$U_q \approx 0.015 \text{ div} .$

Delta Time Measure-ment Accuracy (Full Bandwidth)	Conditions	Accuracy
	Single-shot, Sample mode	±(1 sample interval + 100 ppm × reading + 0.6 ns) (b)
	> 16 averages	±(1 sample interval + 100 ppm × reading + 0.4 ns) (a)
	Sample interval = s/div ÷ 250	

Fig. 1.20 Excerpt from horizontal specifications for TDS1000B–TDS2000B Series oscilloscopes [13]. The specifications considered are marked (**a**), (**b**).

1.8.2   Horizontal (Time) Measurements

When measuring with a digital oscilloscope the time instant at which a certain event occurs (for example, a rising edge), the quantization step Q is equal to the sampling interval T_s. Therefore, in time measurements,

$$U_q = \frac{Q}{2} = \frac{T_s}{2}.$$

Clearly, T_s does not denote here the internal time sampling used by the instrument to acquire the signal, but simply the separation between consecutive time measurements on the scope screen. It is therefore the sampling interval "on the screen," which depends on the horizontal settings.

Moreover, with this kind of instrument one is usually interested only in measuring *differences of time instants* with a guaranteed accuracy. Therefore, we must take into consideration, as before, formula (1.60) for the worst-case uncertainty of the difference of measurements $\hat{\theta} = y_1 - y_2$.

An excerpt from the specifications, of interest for us, is presented in Fig. 1.20.

First, we consider specification (a). Since it is for an averaged waveform, we assume that in this case the effect of noise is negligible.[4]

We note that the specification has exactly the same form of (1.60), and the term $2U_q = T_s$ is clearly identifiable. Comparing the specification with (1.60), we have

$U_{r,G} = 100$ ppm,
$U_{inl} = 0.2$ ns,
$U_q = 0.02\% \, x_{FS}$.

The third figure is obtained by considering that the sample interval T_s is equal, according to specifications, to the sweep speed setting, in seconds/div, divided by

---

[4]Time measurements are affected also by "time noise," commonly known as jitter. Jitter affects instrumentation, telecommunication appliances, etc., also in a complex way. Going deeper into this subject is far beyond the scope of this book.

250. There are therefore 250 samples per division, or 2500 samples in the whole scale of 10 divisions. The time sampling is therefore $T_s = x_{FS}/2500 = 0.04\% \, x_{FS}$.

Specification (b) is very similar, with a slightly higher uncertainty. This is arguably due to the fact that measurement is on a single waveform, not on averaged waveforms. Therefore, even if the specification has a worst-case form, it probably includes an uncertainty contribution due to random errors.

Finally, we point out that in dealing with time measurements, especially with oscilloscopes, the terminology may be varied. For example:

- quantization error can be called "discretization error";
- integral nonlinearity can be called "time base distortion";
- gain error can be called "sampling frequency error" or "sweep speed error".

1.8.3 Example of Uncertainty Specifications for the Keysight, InfiniiVision, 2000 X-Series Oscilloscopes

1.8.3.1 Vertical (Voltage) Measurements

The oscilloscopes of this series, too, have a vertical resolution of eight bits. Therefore,

$$U_q = \frac{Q}{2} = \frac{x_{FS}}{512} \cong 0.2\% \, x_{FS} . \tag{1.62}$$

An excerpt from the specifications [14] is shown in Fig. 1.21.

Once again, we note that both specifications of interest, marked (a) and (b), have the form of the WCU of a difference of measurement (see (1.60)) and a direct measurement (see (1.57)), respectively.

Specification (a) is for "dual cursor" measurements, i.e., difference of measurements $\hat{\theta} = y_1 - y_2$. Comparing the specification with (1.60), we obtain

$$2 \left(U_{inl} + U_q \right) = 0.5\% \, x_{FS} , \tag{1.63}$$

and therefore, since $U_q \cong 0.2\% \, x_{FS}$, we have $U_{inl} \cong 0.05\% \, x_{FS}$.

Vertical system analog channels	
DC vertical gain accuracy [1]	± 3% full scale (≥ 10 mV/div); ± 4% full scale (< 10 mV/div) [2]
DC vertical offset accuracy	± 0.1 div ± 2mV ± 1% of offset setting

Cursors	
Cursors [2]	(b) → – Single cursor accuracy: ± [DC vertical gain accuracy + DC vertical offset accuracy + 0.25% full scale] (a) → – Dual cursor accuracy: ± [DC vertical gain accuracy + 0.5% full scale] [1]

Fig. 1.21 Excerpt from specifications for InfiniiVision 2000 X-Series oscilloscopes [14] concerning vertical (voltage) measurements. The considered specifications are marked (**a**), (**b**).

Specification (b) is perfectly consistent with this assignment of values to U_q and U_{inl}. It also contains, of course, the offset uncertainty. Some notes about the offset and the gain uncertainties are appropriate:

- The "DC vertical offset accuracy" includes the term "1% of offset setting," which can be written $1\%|V_P|$ (vertical position). Therefore, also in this case, U_O is different for $V_P = 0$ and $V_P \neq 0$.
- The "DC vertical gain accuracy" is declared "3% full scale," that is, it is recommended to multiply $U_{r,G}$ by x_{FS}. In this way, as mentioned in discussing the specifications for the Tektronix TDS1000B–TDS2000B Series, it is not necessary to write different formulas for the two cases $V_P = 0$ and $V_P \neq 0$.
- The contributions to uncertainty are often separated by the symbol $\pm$. A reasonable interpretation is that they are independent contributions that actually add up only "in the worst case." Below, we write the specifications in terms of the worst case (addition).

To summarize, the specifications can be interpreted as follows:

- $U_{r,G} = 3\%$, for vertical sensitivity $\geq 10\,\text{mV/div}$;
- $U_{r,G} = 4\%$, for vertical sensitivity $< 10\,\text{mV/div}$;
- $U_O = 0.1\,\text{div} + 2\,\text{mV}$, for $V_P = 0$;
- $U_O = 0.1\,\text{div} + 2\,\text{mV} + 1\%|V_P|$, for $V_P \neq 0$;
- $U_{inl} \cong 0.05\,x_{FS}$;
- $U_q \cong 0.2\%\,x_{FS}$.

1.8.3.2   Horizontal (Time) Measurements

An excerpt from the specifications of interest to us is shown in Fig. 1.22.

Again, specification (a) has exactly the same form of (1.60), like that reported in Fig. 1.20, but the term $2U_q = T_s$ is not explicit here. It is clear, however, that the number of discrete-time cursor positions on the screen is 625, and therefore $T_s = Q = x_{FS}/625 = 0.0016\,x_{FS}$. The term "0.0016 screen width" corresponds, therefore, to $2U_q = Q$. By comparing specification (a) with (1.60), we have:

Horizontal system analog channels

		2002A	2004A	2012A	2014A	2022A
Time base range		5 ns/div to 50 s/div				2 ns/d
Time base accuracy [1]		25 ppm ± 5 ppm per year (aging)				
Time base delay time range	Pre-trigger	Greater of 1 screen width or 200 μs (400 μs in interleaving mode)				
	Post-trigger	1 s to 500 s				
Channel-to-channel deskew range		± 100 ns				
Δ Time accuracy (using cursors)		± (time base accuracy [1] reading) ± (0.0016 [1] screen width) ± 100 ps				(a)

Fig. 1.22 Excerpt from specifications for InfiniiVision 2000 X-Series oscilloscopes [14] concerning horizontal (time) measurements. The specification considered is marked (**a**)

- $U_{r,G} = 25$ ppm (it may vary by ± 5 ppm per year);
- $U_{inl} = 50$ ps;
- $U_q = 0.0008\, x_{FS}$.

References

1. Fluke Corporation: 56X Infrared Thermometers Users Manual. August 2010. https://www.fluke.com/en-us/product/temperature-measurement/ir-thermometers/fluke-62-max
2. Joint Committee for Guides in Metrology (JCGM), JCGM 100:2008: Evaluation of Measurement Data - Guide to the Expression of Uncertainty in Measurement (First edn. 2008; Corrected version 2010)
3. JCGM 100:2008: Guide to the Expression of Uncertainty in Measurement. https://www.iso.org/sites/JCGM/GUM/JCGM100/C045315e-html/C045315e.html?csnumber=50461
4. Joint Committee for Guides in Metrology (JCGM), JCGM 200:2012: International Vocabulary of Metrology - Basic and General Concepts and Associated Terms (VIM), 3rd edn. (First edn. 2008; Corrected version 2012)
5. VIM definitions with informative annotations. https://jcgm.bipm.org/vim/en/index.html
6. Joint Committee for Guides in Metrology (JCGM), JCGM 104:2009 - Evaluation of Measurement Data - An Introduction to the "Guide to the Expression of Uncertainty in Measurement" and Related Documents, First edn. July 2009
7. Arpaia P, Baccigalupi C, Martino M (2015) Type-A Worst-Case Uncertainty for Gaussian noise instruments. J Instrum 10(07). https://doi.org/10.1088/1748-0221/10/07/P07007
8. Attivissimo F, Giaquinto N, Savino M (2007) Worst-case uncertainty measurement in ADC-based instruments. Comput Stand Interfaces 29(1):5–10. https://doi.org/10.1016/j.csi.2005.12.002
9. Smith A, Monti A, Ponci F (2009) Uncertainty and worst-case analysis in electrical measurements using polynomial Chaos theory. IEEE Trans Instrum Meas 58(1):58–67. https://doi.org/10.1109/TIM.2008.2004986
10. Fabbiano L, Giaquinto N, Savino M, Vacca G (2016) On the worst case uncertainty and its evaluation. J Instrum 11(02). https://doi.org/10.1088/1748-0221/11/02/P02001
11. IEEE Standard for Digitizing Waveform Recorders. In: IEEE Standard 1057–2017 (Revision of IEEE Standard 1057–2007), 26 Jan 2018. https://doi.org/10.1109/IEEESTD.2018.8291741
12. Widrow B, Kollar I, Liu M-C (1996) Statistical theory of quantization. IEEE Trans Instrum Meas 45(2):353–361. https://doi.org/10.1109/19.492748
13. Tektronix, Inc.: TDS 1000- and TDS 2000-Series Digital Storage Oscilloscope User Manual. 30 Jan 2002. https://www.tek.com/oscilloscope/tds1002-manual/tds1000-and-tds2000-series-user-manual
14. Keysight Technologies: InfiniiVision 2000 X-Series Oscilloscopes - Data Sheet. 8 Aug 2018. https://www.keysight.com/us/en/assets/7018-02733/data-sheets/5990-6618.pdf

Chapter 2
Time-Domain Measurements

Abstract The present chapter presents some practical laboratory experiences related to time-domain measurements. The instrument used for this purpose is the oscilloscope. First a theoretical introduction is provided to understand the basic operating principles. Successively, some didactic laboratory examples are presented to be used as a practical tutorial in using oscilloscope. In particular, the characterization of passive filters and the measurement of parameters of operational amplifiers are described, as well as the evaluation of the related measurement uncertainty according to the theory described in Chap. 1.

2.1 Basic Theory and Functionalities of Oscilloscopes

Oscilloscopes allow to display the waveform of voltage signals and to carry out a number of measurements, such as voltage amplitude and time measurements. On-screen measurements are possible thanks to a grid pattern made of eight vertical divisions and ten horizontal divisions. The divisions, in turn, are divided into five minor divisions. A typical oscilloscope display grid is shown in Fig. 2.1.

As a voltage meter, the input impedance of an oscilloscope is generally very high. It can be represented as an equivalent ohmic-capacitive parallel impedance, as shown in Fig. 2.2. Typically, the resistance value is of $1\,M\Omega$, while the capacitance value depends on the cable attached to the instrument. When the oscilloscope is used with its dedicated 10X probe with frequency compensation (a probe that reduces signal amplitudes by 10), the resistance value at the probe tip is $10\,M\Omega$, and the capacitance is typically of the order of $10–15\,pF$.

Historically, oscilloscopes are classified into analog or digital, depending on the architecture. In the analog oscilloscope, after a preliminary conditioning, the signal is sent to the visualization block. In the digital oscilloscope, the input signal, after conditioning, is converted into the numerical domain (analog to digital conversion), processed and displayed. Differently from analog oscilloscopes, in the digital ones, the signal is sampled (discretized along the time axis) and digitized (discretized along the vertical axis), before the visualization.

© Springer Nature Switzerland AG 2020   41
A. Cataldo et al., *Basic Theory and Laboratory Experiments in Measurement and Instrumentation*, Lecture Notes in Electrical Engineering 663,
https://doi.org/10.1007/978-3-030-46740-1_2

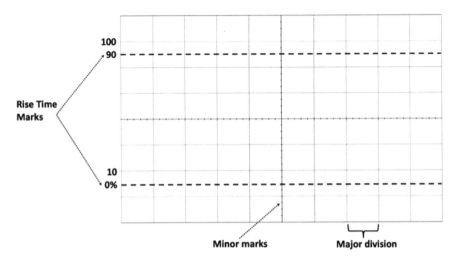

Fig. 2.1 Sketch of the display of an oscilloscope

Fig. 2.2 Ohmic-capacitive
equivalent input impedance,
as seen at the tip of a 10X
probe with frequency
compensation

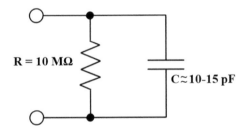

Although the use and commercialization is now limited, to better understand
the working principle of oscilloscopes, it is useful to describe briefly analog oscilloscopes. The operation and display modes of digital oscilloscopes are generally
different from those of the analog oscilloscope; however, the functionalities and the
main commands are substantially the same.

The core of the analog oscilloscope is the visualization system, which is based
on a Cathode Ray Tube (CRT), shown in Fig. 2.3 along with the main components.
In the CRT, the electrostatic deflection of an electron beam is exploited to reproduce
on the screen the temporal evolution of the signal.

A beam of electrons is emitted in an incoherent way from a filament that heats
up, as a result of thermionic effect. These electrons are made coherent and are transformed into a collimated beam, after being accelerated and focused through electrostatic lenses. There is also an horizontal and vertical deflection system, so that the
electronic beam (suitably deflected, along the horizontal and vertical axes) excites
the phosphors that are deposited on the back of the screen. The phosphors become
phosphorescent when they are excited by the incidence of electrons, thus reproducing
the signal waveform. However, the phosphors persist only for a short time, and this

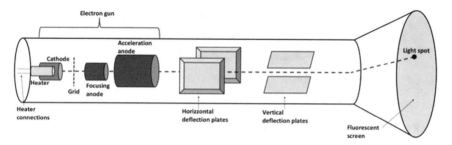

Fig. 2.3 Cathode Ray Tube

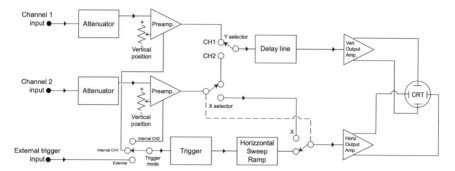

Fig. 2.4 Block diagram of an analog oscilloscope

causes a problem of stability of the trace, depending both on the signal frequency (which determines the luminous intensity of the trace) and on the speed with which the beam goes right and left.

In the deflection system the two pairs of plates are driven by a voltage.

The vertical one is driven by a voltage proportional to the input signal after appropriate conditioning (filtering, attenuation and amplification); so it is faithful to the amplitude variations, as detailed in Fig. 2.4.

The horizontal one is driven by the saw-tooth voltage signal; therefore, it is directly proportional to time. It resets every time it reaches the right side of the screen and starts again.

As shown in Fig. 2.4, the oscilloscope can also have multiple vertical channels to display multiple signals simultaneously. However, the time base system is generally common to all channels, so the time settings are the same for all the vertical channels. If the signals differ greatly in frequency, they cannot be suitably displayed simultaneously (further details can be found in [1]).

Differently from analog oscilloscopes, DSOs provide permanent signal storage and extensive waveform processing possibilities. Some of the DSOs subsystems are similar to those in analog oscilloscopes. A DSO employs a serial-processing architecture to acquire, process and display a signal waveform on its screen, as shown in Fig. 2.5.

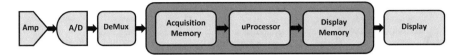

Fig. 2.5 The serial-processing architecture of a digital storage oscilloscope (DSO)

Digital oscilloscopes use an analog-to-digital converter (ADC) to convert the measured voltage into digital information. DSO acquire the waveform as a series of samples, and stores these samples until it accumulates enough of them to describe a waveform. The digital oscilloscope then re-assembles the waveform for display on the screen. The digital approach means that the oscilloscope can display a signal waveform (with frequency in the oscilloscope working frequency range) with good stability, brightness, and clarity (the brightness and stability of the trace do not depend on the local signal frequency or speed). For repetitive signals, the bandwidth of the digital oscilloscope is a function of the analog bandwidth of the front-end components of the oscilloscope, commonly referred to as the −3 dB point. For single-shot and transient events, such as pulses and steps, the bandwidth can be limited by the oscilloscope's sample rate.

Similarly to analog oscilloscopes, the first stage of a DSO is an initial conditioning block (attenuation or amplification, low-pass filtering). This block adapts the input signal to the following blocks. Vertical controls allow to adjust the amplitude and position range at this stage. Next, the ADC block in the horizontal system samples the analog signal at equidistant time instants, and converts the amplitude of the signal into discrete digital values, the signal samples.

2.1.1 Main Functionalities

Typically, for analog or digital versions, the front panel of an oscilloscope is divided into three main sections, called the vertical, horizontal, and trigger sections.

Accordingly, in using an oscilloscope, one needs to adjust the following major settings for a correct signal visualization:

- Attenuation or amplification of the signal: using the volts/div control (k_v) allows one to adjust the amplitude measurement range, adapting it to the amplitude of the measured signal.
- Time base: using the seconds/div or time/div control (k_t) allows one to set the duration of each horizontal division for displaying the desired total time interval.
- Trigger: using the trigger level allows one to stabilize the visualization of a periodic signal waveform or trigger on a single event.

Vertical controls can be used to position and scale the waveform vertically. These can also be used to set the input coupling and other signal conditioning.

It is worth noting that an important setting of the oscilloscope is the signal coupling. The coupling can be set to DC, AC, or ground. DC coupling shows both

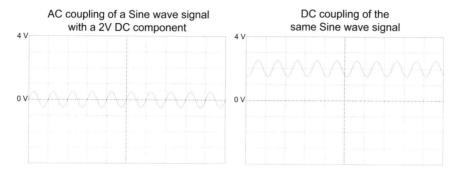

Fig. 2.6 AC and DC input coupling

the continuous and alternate components of the input signal, while AC eliminates the continuous (i.e., a high-pass filtering excludes the signal frequency components around the zero frequency). Figure 2.6 shows the difference between the two coupling modalities. The ground setting disconnects the input signal from the vertical system, which shows the position of the zero-volt level on the display.

One of the major advantages of digital oscilloscopes is their ability to store waveforms. It is also possible to start/stop the acquisition system as desired by the user. This process can be done automatically by the oscilloscope, which stops acquiring after one acquisition is complete or after one set of records has been turned into an envelope or average waveform. This feature is commonly called single sweep or single sequence or single shot.

With regard to the trigger system functionalities, they are necessary for synchronizing the horizontal sweep with a suitable point of the signal. Trigger controls allow one to stabilize repetitive waveforms and capture single-shot waveforms. The trigger makes repetitive waveforms appear stable on the oscilloscope display by repeatedly displaying the same portion of the input signal. Fixing the trigger signal is important to avoid the situation shown in Fig. 2.7. It is possible to find advanced trigger controls (according to the model of the oscilloscope) that enable one to isolate specific events of interest to optimize the oscilloscope's sample rate and record length.

Horizontal trigger position control is available only on digital oscilloscopes. It actually represents the horizontal position of the trigger in the waveform record. Varying the horizontal trigger position allows one to capture what a signal did before a trigger event, known as pretrigger viewing. Thus, it determines the length of the viewable signal both preceding and following a trigger point. Digital oscilloscopes can provide pretrigger viewing because they constantly sample the input signal, whether or not a trigger has been received.

The trigger level and slope controls provide the basic trigger point definition and determine how a waveform is displayed, as illustrated in Fig. 2.8. The slope control determines whether the trigger point is on the rising or falling edge of a signal. A rising edge is a positive slope, and a falling edge is a negative slope. The level control determines the amplitude at which the signal is triggered.

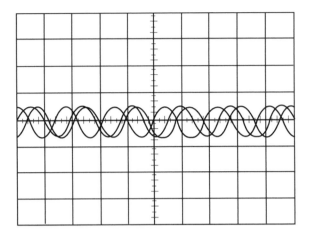

Fig. 2.7 Example of the visualization of an untriggered signal waveform

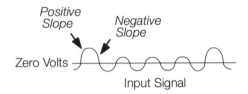

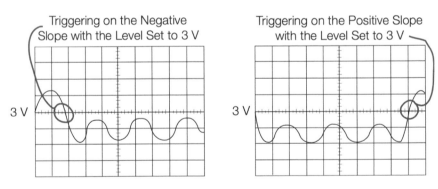

Fig. 2.8 Example of signal triggering: negative slope (left) and positive slope triggering (right)

The oscilloscope does not necessarily need to trigger on the signal being displayed. In fact, many different triggers can be selected, such as:

- channel 1, 2, etc.;
- "ext" input, where an external signal other than that applied to input channels is applied;
- "line" or "power" input, which uses the power source as the signal for triggering;

Common trigger *modes* include normal and auto. In normal mode, the oscilloscope sweeps only if a trigger event occurs. In auto mode, after waiting in vain for a trigger event, for an internally prefixed time, the oscilloscope sweeps and presents the obtained waveform. If no signal is present, a timer in the oscilloscope triggers the sweep. This ensures that the display will not disappear if the signal does not cause a trigger event. Additionally, many oscilloscopes include advanced trigger modalities such as single-sweep triggering, triggering on video signals, and automatically setting the trigger level.

2.1.2 Sampling Modalities

As mentioned previously, the measurement signal is sampled through the sample&hold circuit (more details can be found in [2]). Theoretically, the sampling frequency, f_S, must satisfy the *Nyquist–Shannon theorem*:

$$f_S > 2f_{max} , \tag{2.1}$$

where f_{max} is the highest frequency component of the signal to be sampled. In practice, to obtain a satisfactory visualization of the signal, a suitable value of the sampling frequency is $f_S \simeq 10 f_{max}$.

An inherent operation in A-to-D conversion is quantization, discussed theoretically in Chap. 1. During quantization, the sampled signal is associated with defined levels. The number of bits, b, used to represent the analog signal determines the ADC resolution:

$$LSB = Q = \frac{x_{FS}}{2^b} , \tag{2.2}$$

where FS is the interval between the maximum and minimum levels of the analog signal that the ADC can digitize. The LSB (Least Significant Bit) theoretically identifies the smallest variation of the analog input signal that can be detected.

The sampling process is managed by the time base, while the trigger section allows one to obtain the synchronization necessary for a correct display of the acquired signal.

One of the major advantages offered by digital oscilloscopes relies on the possibility of resorting to two sampling modalities:

• real-time sampling;
• equivalent-time sampling.

Real-time sampling is suitable for signals whose frequency range is lower than half the oscilloscope's maximum sample rate. In this case, in fact, the oscilloscope can acquire a number of points in one "sweep" sufficient to reconstruct an accurate waveform. This kind of sampling is shown in Fig. 2.9.

Equivalent-time sampling (which, in turn, can be categorized as random or sequential) offers the advantage of correctly acquiring and visualizing a periodic or repetitive signal whose frequency content can be much higher than that required by the sampling theorem.

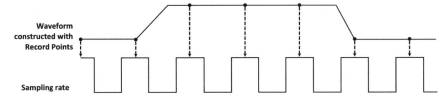

Fig. 2.9 Real-time sampling

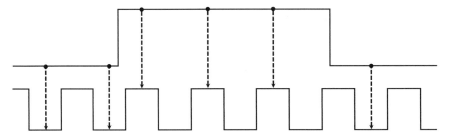

Fig. 2.10 Sequential equivalent-time sampling

It goes without saying that when fast or single-shot or transient signals must be acquired (i.e., aperiodic signals), real-time sampling becomes mandatory. If the sample rate is not fast enough, high-frequency components can "fold down" into a lower frequency, causing aliasing. In addition, real-time sampling is further complicated by the high-speed memory required to store the waveform once it is digitized.

Equivalent-time sampling overcomes this problem and hence can be used to accurately acquire signals whose frequency exceeds half the oscilloscope's sample rate, provided that the signal is periodic. The strategy behind equivalent-time sampling is to sample points in different repetitions of the signal.

The sequential equivalent-time sampling acquires one sample per trigger, independently of the time/div setting, or sweep speed, as illustrated in Fig. 2.10.

When a trigger is detected, a sample is taken after a very short delay. When the next trigger occurs, a small time increment, Δt, is added to this delay, and the digitizer takes another sample. This process is repeated many times, with "Δt" added to each previous acquisition, until the time window is filled. Sample points are ordered and displayed as if they belonged to a single repetition.

Random equivalent-time digitizers (samplers) utilize an internal clock that runs asynchronously with respect to the input signal and the signal trigger, as shown in Fig. 2.11.

Samples are taken continuously, independently of the trigger position, and are displayed based on the time difference between the sample and the trigger. Although samples are taken sequentially in time, they are random with respect to the trigger. Sample points appear randomly along the waveform when displayed on the oscilloscope screen. This kind of equivalent-time sampling allows one also to display portions of the signal prior to the trigger event. This is the pretrigger modality.

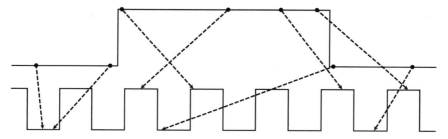

Fig. 2.11 Random equivalent-time sampling

2.2 Didactic Experiments Involving Time-Domain Measurements: Uncertainty Evaluation

The following sections address a set of didactic laboratory experiments that require time-domain measurements using the DSO and proper mathematical operations involving the measurement results (indirect measurements). It is therefore useful to make some considerations valid for all the experiments, about the computations required for uncertainty evaluations. The list of the employed equipment, in all the experiments, is the following.

> **Equipment:**
> 1. A digital oscilloscope (Keysight InfiniiVision DSO X-2012A [3])
> 2. A PCB with a filter circuit or an op-amp circuit (example circuit diagrams are reported in Appendix B)
> 3. A function generator (Hewlett-Packard 33120A)
> 4. A power supply unit ±15 V (Hewlett-Packard E3631A—used only for op-amp experiments [6])
> 5. Connection cables

The uncertainty evaluation is of *type B* and relies on the metrological specifications of the instruments. Furthermore, the *worst-case uncertainty* discussed in Chap. 1, or WCU, is evaluated, and the related propagation rules are used. This choice is consistent with the metrological specifications of the instruments, which are in terms of WCU. Moreover, it is justified by the fact that in the considered indirect measurements, typically only two quantities measured by the instruments are involved, which is a small number. In a single case, three quantities are involved, i.e., one peak-to-peak voltage measurement and two resistance values.

It is perfectly possible to convert the instrument's specifications into standard uncertainties and propagate them using the related rules, as illustrated in Chap. 1. In this way, one does not stick closely to the instrument's specifications, but the operations are more GUM-compliant. The interested reader can verify that for the cases

considered here, the final GUM-compliant uncertainty values with 95% coverage
probability are smaller than the WCU computed here, but the values are quite close.
Therefore, using worst-case uncertainties for the presented cases is a conservative
choice, but not incorrect or overly pessimistic.

2.2.1 Uncertainty of Measurements Provided by Instruments

The measurements provided by the instruments and used in the experiments in this
chapter are the following:

- peak-to-peak voltages V, measured by the DSO;
- time delays Δt between sinusoidal waveforms, measured by the DSO;
- frequency f of sinusoidal waveforms, measured by the function generator.

For peak-to-peak voltages V, the instrument's metrological specifications are reported
in Fig. 2.12.

Peak-to-peak voltages $V_{max} - V_{min} = \max(v(t)) - \min(v(t))$ are "differences of
measurements" of the form $\theta = x_1 - x_2$ discussed in Sect. 1.7.3.2, and their WCU
is of the form

$$U(\hat{\theta}) = U_{r,G} \cdot x_{FS} + 2(U_{inl} + U_q). \qquad (2.3)$$

This expression, as discussed in Sect. 1.8.1.1, multiplies the worst-case gain uncertainty $U_{r,G}$ by the full-scale range x_{FS}, so that the effect of the vertical position setting
is automatically taken into account. The measurement, as highlighted in Sect. 1.7.3.2,
is not affected by the offset error of the instrument (unlike a measurement of absolute
voltage, such as, e.g., $V_{max} = max(v(t))$. The WCU is calculated by applying the
specification for "dual cursor accuracy" highlighted in Fig. 2.12. In order to evaluate the uncertainty for actual measurements, it is important to remember that if we
denote by k_V the vertical sensitivity in volts/div, the full-scale range is $x_{FS} = 8 \cdot k_V$,
since there are eight vertical divisions. Therefore, the vertical sensitivity k_V must

DC vertical gain accuracy*	±3% full scale (≥ 10 mV/div); ±4% full scale (< 10 mV/div) **
Cursors**	• Single cursor accuracy: ±[DC vertical gain accuracy + DC vertical offset accuracy + 0.25% full scale] • Dual cursor accuracy: ±[DC vertical gain accuracy + 0.5% full scale]*

* Denotes warranted specifications, all others are typical.
  Specifications are valid after a 30-minute warm-up period and from ±10 °C firmware calibration temperature.
** 1 mV/div and 2 mV/div is a magnification of 4 mV/div setting. For vertical accuracy calculations, use full scale
   of 32 mV for 1 mV/div and 2 mV/div sensitivity setting.

Fig. 2.12 Uncertainty specifications for the vertical channel of the Keysight InfiniiVision DSO
X-2012A oscilloscope

Time base accuracy*	25 ppm ±5 ppm per year (aging)
Δ Time accuracy (using cursors)	± (time base accuracy* reading) ± (0.0016 * screen width) ± 100 ps

* Denotes warranted specifications, all others are typical.
Specifications are valid after a 30-minute warm-up period and from ±10 °C firmware calibration temperature.

Fig. 2.13 Time-based accuracy and Δ time accuracy

FREQUENCY CHARACTERISTICS

Sine:	100 μHz – 15 MHz
Square:	100 μHz – 15 MHz
Triangle:	100 μHz – 100 kHz
Ramp:	100 μHz – 100 kHz
Noise (Gaussian):	10 MHz bandwidth
Arbitrary Waveforms:	
8 to 8,192 points:	100 μHz – 5 MHz
8,193 to 12,287 points:	100 μHz – 2.5 MHz
12,288 to 16,000 points:	100 μHz – 200 kHz
Resolution:	10 μHz or 10 digits
Accuracy:	10 ppm in 90 days, 20 ppm in 1 year, 18°C – 28°C
Temperature Coefficient:	<2 ppm / °C
Aging:	< 10 ppm / yr

OUTPUT CHARACTERISTICS [1]

Amplitude (into 50Ω): [2]	50 mVpp – 10 Vpp	
Accuracy (at 1 kHz):	± 1% of specified output	
Flatness	(sine wave relative to 1 kHz)	
< 100 kHz:	± 1% (0.1 dB)	
100 kHz to 1 MHz:	± 1.5% (0.15 dB)	
1 MHz to 15 MHz:	± 2% (0.2 dB) Ampl ± 3Vrms	
1 MHz to 15 MHz:	± 3.5% (0.3 dB) Ampl < 3Vrms	
Offset (into 50Ω): [3]	± 5 Vpk ac + dc	
Accuracy: [4]	± 2% of setting + 2 mV	
Output Impedance:	50 ohms fixed	
Resolution:	3 digits, Amplitude and Offset	
Output Units:	Vpp, Vrms, dBm	
Isolation:	42 Vpk maximum to earth	

Fig. 2.14 Function generator datasheet

always be reported together with the measured value. Time delays Δt are also "differences of measurements." The related metrological specifications of the instrument are reported in Fig. 2.13. Like those for voltage measurements, they have been discussed and interpreted in Sect. 1.8.3.

The worst-case uncertainty of $\hat{\theta} = \Delta t$ is calculated by applying the specification for "Δ time accuracy" highlighted in Fig. 2.13, which is of the form

$$U(\hat{\theta}) = U_{r,G} \cdot |\hat{\theta}| + 2(U_{inl} + U_q). \tag{2.4}$$

As for voltages, it is important to note that in order to compute the uncertainty for actual measurements, one must take note of the sweep speed k_t in seconds/div and compute the full-scale range $x_{FS} = 10 \cdot k_t$, since there are 10 horizontal divisions.

The value of the frequency f is obtained by the signal generator setting: the datasheet of the instrument is shown in Fig. 2.14. According to the datasheet, the relative worst-case uncertainty in the frequency measurement is $U_r(f) = 20\,\text{ppm}$, and the absolute uncertainty is therefore $U_r(f) \cdot |f|$.

2.2.2 Uncertainty of Computed Measurements in Experiments

In almost all the experiments, we compute the ratio of two peak-to-peak voltages, of the kind

$$A = V_1/V_2 . \tag{2.5}$$

For example, V_1 is the output of a filter, V_2 the input, and A its magnitude response. In this computation we consider the errors affecting V_1 and V_2 to be independent. This is justified by the fact that they are measured by two separate channels of the DSO, and therefore they are not necessarily affected by the same gain and offset errors (as would be the case with V_{max} and V_{min} measured on the same waveform). By applying the LPU for worst-case uncertainties with independent errors, we have

$$U_r(A) = U_r(V_1) + U_r(V_2) , \tag{2.6}$$

where, of course, $U_r(V) = U(V)/|V|$ and $U(A) = U_r(A) \cdot |A|$. We compute also the value of A in decibel units, i.e.,

$$A_{dB} = 20 \cdot \log_{10}(A) . \tag{2.7}$$

The absolute uncertainty of this quantity, using the LPU, is given by

$$U(A_{dB}) = |c_A| \cdot U(A) , \tag{2.8}$$

where the absolute sensitivity coefficient c_A is

$$c_A = \frac{\partial A_{dB}}{\partial A} = 20 \cdot \frac{\partial}{\partial A} \cdot \frac{\ln(A)}{\ln(10)} = \frac{20}{\ln(10)} \cdot \frac{1}{A} . \tag{2.9}$$

Consequently,

$$U(A_{dB}) = \frac{20}{\ln(10)} \cdot \frac{U(A)}{|A|} = \frac{20}{\ln(10)} \cdot U_r(A) . \tag{2.10}$$

It is interesting to note that the absolute uncertainty $U(A_{dB})$ is a scaled version of the relative uncertainty $U_r(A) = U_r(V_1) + U_r(V_2)$. Indeed, this absolute uncertainty defines the maximum *relative* variation that can be attributed to the values of A. For this reason, computing an absolute uncertainty of A_{dB} is appropriate, while computing a relative uncertainty $U_r(A_{dB}) = U(A_{dB})/A_{dB}$ makes little sense.

Another frequent computation is the phase shift (in degrees) between sinusoidal waveforms:

$$\Delta\phi = -360° \cdot f \cdot \Delta t . \tag{2.11}$$

A clarification about the minus sign in this equation and the signs of Δt and $\Delta \phi$ is needed. If $v_1(t)$ and $v_2(t)$ are two sinusoidal waveforms $v_i(t) = V_i \cdot sin(2\pi f t + \phi_i)$, $i = 1, 2$, we define $t_i = -\phi_i/(2\pi f)$, the zero-crossing instants of the waveform, corresponding to the condition $2\pi f t + \phi_i = 0$. With this definition, the time shift $\Delta t = t_1 - t_2$ and the phase shift $\Delta \phi = \phi_1 - \phi_2$ are linked by the relation (2.11). For example, if $v_1(t)$ is the output of an RC filter and $v_2(t)$ the input, so that $\phi_2 < \phi_1$, we will measure a positive Δt and a negative $\Delta \phi$.

Since f and Δt are measured by different instruments (the function generator and the DSO, respectively), we consider the errors independent, and according to the propagation law, we have

$$U_r(\Delta \phi) = U_r(f) + U_r(\Delta t), \qquad (2.12)$$

and, of course, $U(\Delta \phi) = U_r(\Delta \phi)|\Delta \phi|$.

Finally, in order to measure the open-loop gain in an operational amplifier, we will have to consider the uncertainty propagation for a voltage divider,

$$V_d = V' \cdot \frac{R_4}{R_3 + R_4}. \qquad (2.13)$$

For this expression, it is simpler to consider the propagation law for relative quantities. For an indirect measurement $\theta = f(x_1, \ldots, x_N)$ (true values), $\hat{\theta} = f(y_1, \ldots, y_N)$ (measured values), this propagation is expressed by Eqs. (1.31) and (1.34), which we rewrite here:

$$c_{r,n} = \frac{\partial \ln(\theta)}{\partial x_n} \cdot x_n,$$

$$U_r(\hat{\theta}) = \sum_{n=1}^{N} |c_{r,n}| U_r(y_n).$$

For the voltage divider, the result is

$$U_r(V_d) = U_r(V') + \left| \frac{R_3}{R_3 + R_4} \right| [U_r(R_3) + U_r(R_4)]. \qquad (2.14)$$

In the appropriate section, we will evaluate this expression assuming $U_r(R_3) = U_r(R_4) = 5\%$ (the tolerance of the resistors used). Of course, the uncertainty can be made much smaller by measuring the resistors with an ohmmeter. Other uncertainty formulas used in this Chapter are easy to derive, and left to the reader as a useful exercise.

As a final note, we highlight that in the following, uncertainties like $U(y)$, $U_r(y)$ will be denoted using the slightly more compact alternative form U_y, $U_{r,y}$.

2.3   Didactic Laboratory Experiments on Passive Filters

The goal of this section is to describe practical measurements on passive filters, so that the real behavior of the considered circuits is compared with their ideal model. The frequency responses of several filters are measured experimentally, and the corresponding Bode diagrams are derived, along with the related uncertainties.

2.3.1   General Measurement Procedure

To measure the frequency response at a given frequency f, a sinusoidal signal is generated by a function generator and applied at the filter input. The signals at the input and at the output of the filter are acquired by two separate channels of a DSO, so that their amplitudes and time delay can be measured. From these data, the attenuation and phase shift values are obtained and then reported on Bode diagrams representing the magnitude and phase responses. The procedural steps in all the experiments are as follows:

1. Connect the function generator output to the "INPUT" port of the PCB (described in Appendix B), with a coaxial cable with BNC-BNC connectors.
2. Connect "OUTPUT1" and "OUTPUT2" of the PCB, respectively, to "CH1" and "CH2" of the oscilloscope, using BNC-BNC connectors. "CH1" of the oscilloscope will show the input signal of the filter, whereas "CH2" will show the output signal.
3. Set manually the attenuation factor of the probe of "CH1" and "CH2" to 1:1 (short BNC-BNC connectors are used instead of attenuation probes; therefore, the probe factor must be 1:1).
4. Select a sinusoidal signal on the function generator and set a starting frequency value (f).
5. Get the automatic measurement of the peak-to-peak voltage for both the "CH1" and "CH2" channels of the DSO.
6. Get the automatic measurement of the delay between the "CH1" and "CH2" channels of the DSO.
7. Vary the frequency f of the sinusoidal signal on the function generator, and repeat steps 6 and 7, to obtain several measurements, which will be useful for plotting the experimental Bode diagrams.

2.3.2   Characterization of an RC Filter

A passive RC filter consists of a capacitor and a resistor connected in series. Figure 2.15 shows the circuital schematization of the filter and a sketch of the exper-

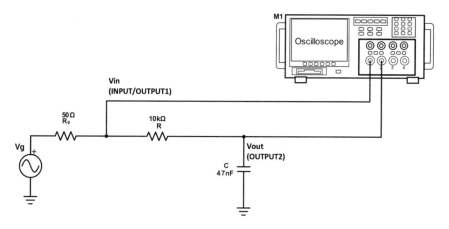

Fig. 2.15 Sketch of the experimental setup for the characterization of an RC filter

imental setup used for its characterization. The values of the resistance (R) and of the capacitance (C) are

- $R = 10\,k\Omega$;
- $C = 47\,nF$.

All the passive components in the circuit have an uncertainty of 5% (i.e., the tolerance of the reference value). These data, as mentioned above, are useful for calculating uncertainties.

This circuit behaves as a low-pass filter, which attenuates the signals with frequency higher than the cutoff frequency (f_c).

The filter transfer function, $H(j\omega)$, can be obtained by applying the voltage divider formula to the circuit. The transfer function is then used to plot the theoretical Bode diagrams and then to evaluate the amplitude and phase of the filter:

$$H(j\omega) = \frac{1}{1 + j\omega RC}. \tag{2.15}$$

The pole is calculated as follows:

$$1 + j\omega CR = 0 \Rightarrow \omega_p = \frac{1}{RC}. \tag{2.16}$$

The cutoff frequency is

$$f_c = \frac{\omega_p}{2\pi} = \frac{1}{2\pi RC}. \tag{2.17}$$

Considering the nominal values of the R and C components, the cutoff frequency can be calculated by applying (2.17): $f_c \simeq 338.6275\,Hz$.

The general measurement procedure described in Sect. 2.3.1 has been applied. Table 2.1 summarizes the measurement results. Amplitude uncertainties are reported in Table 2.2, while time and phase shift uncertainties are summarized in Table 2.3.

Table 2.1 Measurement data for the RC filter

f (Hz)	V_i (V)	k_{V_i} (V/div)	V_o (V)	k_{V_o} (V/div)	Δt (μs)	k_t (μs/div)
10	10.20	2	10.10	2	360.0	500
25	10.20	2	10.10	2	440.0	500
50	10.20	2	10.10	2	448.0	100
75	10.20	2	9.90	2	448.00	50
100	10.20	2	9.70	2	450.00	50
150	10.20	2	9.30	2	426.00	50
200	10.20	2	8.80	2	412.00	50
250	10.20	2	8.20	2	396.00	50
300	10.20	2	7.56	1	374.00	50
339	**10.20**	**2**	**7.12**	**1**	**359.00**	**50**
400	10.20	2	6.51	1	338.00	50
500	10.10	2	5.67	1	306.00	50
600	10.00	2	4.90	1	277.00	50
800	9.60	2	3.74	0.5	230.00	50
1000	9.10	2	2.97	0.5	196.00	50
1500	9.70	2	2.17	0.5	142.40	20
2000	10.10	2	1.73	0.5	111.00	20
5000	10.20	2	0.695	0.1	47.00	10
10000	10.20	2	0.358	0.05	24.20	10

Table 2.2 Evaluated uncertainty values for voltage amplitude measurements in an RC filter, for each considered frequency value. The calculated values for A and the related uncertainties are also reported

f (Hz)	U_{Vi} (V)	$U_{r,Vi}$ (%)	U_{Vo} (V)	$U_{r,Vo}$ (%)	A_{dB} (dB)	$U_{A_{dB}}$ (dB)	A	U_A	$U_{r,A}$ (%)
10	0.56	5.5	0.56	5.5	−0.09	0.96	0.99	0.11	11
25	0.56	5.5	0.56	5.5	−0.09	0.96	0.99	0.11	11
50	0.56	5.5	0.56	5.5	−0.09	0.96	0.99	0.11	11
75	0.56	5.5	0.56	5.7	−0.26	0.97	0.97	0.11	11
100	0.56	5.5	0.56	5.8	−0.44	0.98	0.95	0.11	11
150	0.56	5.5	0.56	6.0	−0.8	1.0	0.91	0.10	12
200	0.56	5.5	0.56	6.4	−1.3	1.0	0.86	0.10	12

(continued)

Table 2.2   (continued)

f (Hz)	U_{Vi} (V)	$U_{r,Vi}$ (%)	U_{Vo} (V)	$U_{r,Vo}$ (%)	A_{dB} (dB)	$U_{A_{dB}}$ (dB)	A	U_A	$U_{r,A}$ (%)
250	0.56	5.5	0.56	6.8	−1.9	1.1	0.80	0.10	12
300	0.56	5.5	0.28	3.7	−2.60	0.80	0.741	0.068	9.2
339	0.56	5.5	0.28	3.9	−3.12	0.82	0.698	0.066	9.4
400	0.56	5.5	0.28	4.3	−3.90	0.85	0.638	0.062	10
500	0.56	5.5	0.28	4.9	−5.01	0.91	0.561	0.059	10
600	0.56	5.6	0.28	5.7	−6.20	0.98	0.490	0.055	11
800	0.56	5.8	0.14	3.7	−8.19	0.83	0.390	0.037	10
1000	0.56	6.2	0.14	4.7	−9.73	0.94	0.326	0.035	11
1500	0.56	5.8	0.14	6.5	−13.0	1.1	0.224	0.027	12
2000	0.56	5.5	0.14	8.1	−15.3	1.2	0.171	0.023	14
5000	0.56	5.5	0.028	4.0	−23.25	0.83	0.0688	0.0066	10
10000	0.56	5.5	0.014	3.9	−29.09	0.82	0.0351	0.0033	9.4

Table 2.3   Evaluated uncertainty values for time and phase shift measurements in an RC filter, for each considered frequency value. The calculated values for $\Delta\phi$ and the related uncertainties are also reported

f (Hz)	$U_{\Delta t}$ (μs)	$U_{r,\Delta t}$ (%)	U_f (Hz)	$\Delta\phi$ (deg)	$U_{\Delta\phi}$ (deg)	$U_{r,\Delta\phi}$ (%)
10	8.0	2.2	0.00020	−1.296	0.029	2.2
25	8.0	1.8	0.00050	−3.960	0.072	1.8
50	1.6	0.36	0.0010	−8.064	0.029	0.36
75	0.81	0.18	0.0015	−12.096	0.022	0.18
100	0.81	0.18	0.0020	−16.200	0.030	0.18
150	0.81	0.19	0.0030	−23.004	0.044	0.19
200	0.81	0.20	0.0040	−29.664	0.059	0.20
250	0.81	0.20	0.0050	−35.640	0.074	0.21
300	0.81	0.22	0.0060	−40.392	0.088	0.22
339	0.81	0.23	0.0068	−43.81	0.10	0.23
400	0.81	0.24	0.0080	−48.67	0.12	0.24
500	0.81	0.26	0.010	−55.08	0.15	0.27
600	0.81	0.29	0.012	−59.83	0.18	0.29
800	0.81	0.35	0.016	−66.24	0.23	0.35
1000	0.80	0.41	0.020	−70.56	0.29	0.41
1500	0.32	0.23	0.030	−76.90	0.18	0.23
2000	0.32	0.29	0.040	−79.92	0.23	0.29
5000	0.16	0.34	0.10	−84.60	0.29	0.34
10000	0.16	0.66	0.20	−87.12	0.58	0.67

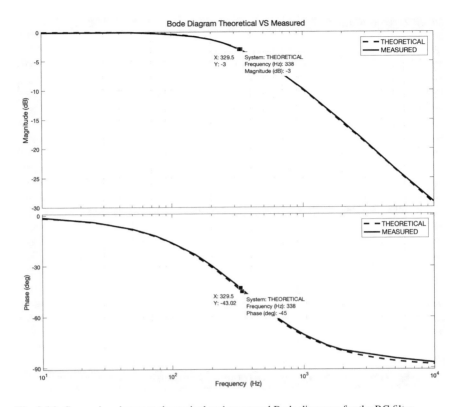

Fig. 2.16 Comparison between theoretical and measured Bode diagrams for the RC filter

In conclusion, it is possible to graphically compare the Bode diagrams of the measured data and the theoretical data, as shown in Fig. 2.16.

2.3.3 Characterization of a CR Filter

The passive CR filter is a circuit consisting of a resistor and a capacitor in series, as shown in Fig. 2.17.

This circuit acts as a high-pass filter, which attenuates signals with frequency lower than the design cutoff frequency (f_c).

The values of the circuital components used for the CR filter are:

- $R = 10 \, k\Omega$;
- $C = 220 \, nF$.

The transfer function is obtained by applying the voltage divider formula to the circuit. The obtained transfer function is then used to plot the theoretical Bode diagrams and to evaluate attenuation and phase variation of the filter:

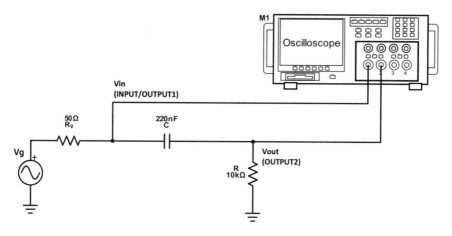

Fig. 2.17 Sketch of the experimental setup for the characterization of a CR filter

$$H(j\omega) = \frac{j\omega RC}{1 + j\omega RC} . \tag{2.18}$$

It has a zero at the origin and a pole that is given by

$$1 + j\omega CR = 0 \Rightarrow \omega_p = \frac{1}{RC} . \tag{2.19}$$

The cutoff frequency is:

$$f_c = \frac{\omega_p}{2\pi} = \frac{1}{2\pi RC} . \tag{2.20}$$

Considering the nominal values of R and C components, the cutoff frequency can be calculated by applying (2.20): $f_c \simeq 72.3432\,\text{Hz}$.

The general measurement procedure described in Sect. 2.3.1 has been applied. Table 2.4 summarizes the measurement results. Amplitude uncertainties are reported in Table 2.5. Time and phase shift uncertainties are reported in Table 2.6.

In conclusion, it is possible to graphically compare the Bode diagrams of the measured data and the theoretical ones, as shown in Fig. 2.18.

2.3.4 Characterization of a CR-RC Filter

In this section we characterize a filter that is a combination of a low-pass filter and a high-pass filter, as shown in Fig. 2.19. With a proper choice of the cutoff frequencies, the result is a very simple band-pass filter, i.e., a device that passes frequencies within a certain range and attenuates frequencies outside that range. The ideal band-

Table 2.4 Measurement data for the CR filter

f (Hz)	V_i (V)	k_{V_i} (V/div)	V_o (V)	k_{V_o} (V/div)	Δt (μs)	k_t (μs/div)
5	10.20	2	0.708	0.1	−47200	10000
10	10.30	2	1.400	0.2	−22700	5000
20	10.20	2	2.73	0.5	−10200	2000
30	10.20	2	3.90	0.5	−6300	1000
50	10.20	2	5.79	1	−3100.0	500
72	**10.20**	**2**	**7.16**	**1**	**−1750.0**	**200**
100	10.20	2	8.30	2	−996.0	200
250	10.20	2	9.80	2	−179.00	50
500	10.10	2	10.00	2	−46.00	10
1000	9.10	2	9.10	2	−11.000	5
5000	10.10	2	10.10	2	−0.360	1

Table 2.5 Evaluated uncertainty values for voltage amplitude measurements in a CR filter, for each considered frequency value. The calculated values for A and the related uncertainties are also reported

f (Hz)	U_{Vi} (V)	$U_{r,Vi}$ (%)	U_{Vo} (V)	$U_{r,Vo}$ (%)	A_{dB} (dB)	$U_{A_{dB}}$ (dB)	A	U_A	$U_{r,A}$ (%)
5	0.56	5.5	0.028	4.0	−23.17	0.82	0.0694	0.0066	9.4
10	0.56	5.4	0.056	4.0	−17.33	0.82	0.136	0.013	9.4
20	0.56	5.5	0.14	5.1	−11.45	0.92	0.268	0.028	11
30	0.56	5.5	0.14	3.6	−8.35	0.79	0.382	0.035	9.1
50	0.56	5.5	0.28	4.8	−4.92	0.90	0.568	0.059	10
72	0.56	5.5	0.28	3.9	−3.07	0.82	0.702	0.066	9.4
100	0.56	5.5	0.56	6.7	−1.8	1.1	0.81	0.10	12
250	0.56	5.5	0.56	5.7	−0.35	0.97	0.96	0.11	11
500	0.56	5.5	0.56	5.6	−0.09	0.97	0.99	0.11	11
1000	0.56	6.2	0.56	6.2	0.0	1.1	1.00	0.12	12
5000	0.56	5.5	0.56	5.5	0.0	1.0	1.00	0.11	11

pass filter has a perfectly flat passband, without either attenuation or gain for the frequencies inside, and completely eliminates all frequencies outside this range. In practice, no band-pass filter is ideal. In particular, no real filter can completely attenuate all the frequencies outside the desired band; there is a region adjacent to the passband where the frequencies are attenuated but not completely. These regions are called "roll-off" regions.

The values of the electrical components used for implementing this filter are:

- $R_1 = 100\,k\Omega$;
- $C_1 = 1\,nF$;

Table 2.6 Evaluated uncertainty values for time and phase shift measurements in a CR filter, for each considered frequency value. The calculated values for $\Delta\phi$ and the related uncertainties are also reported

f (Hz)	$U_{\Delta t}$ (μs)	$U_{r,\Delta t}$ (%)	U_f (Hz)	$\Delta\phi$ (deg)	$U_{\Delta\phi}$ (deg)	$U_{r,\Delta\phi}$ (%)
5	161	0.34	0.00010	84.96	0.29	0.34
10	81	0.35	0.00020	81.72	0.29	0.36
20	32	0.32	0.00040	73.44	0.23	0.32
30	16	0.26	0.00060	68.04	0.18	0.26
50	8.1	0.26	0.0010	55.8	0.15	0.26
72	3.2	0.19	0.0014	45.36	0.085	0.19
100	3.2	0.32	0.0020	35.856	0.12	0.33
250	0.80	0.45	0.0050	16.11	0.073	0.45
500	0.16	0.35	0.010	8.28	0.029	0.35
1000	0.080	0.73	0.020	3.96	0.029	0.73
5000	0.016	4.4	0.10	0.648	0.029	4.4

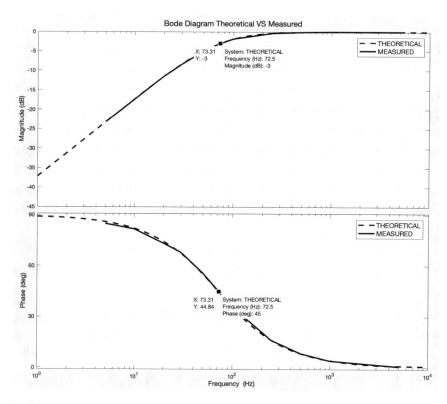

Fig. 2.18 Comparison between theoretical and measured Bode diagrams for the CR filter

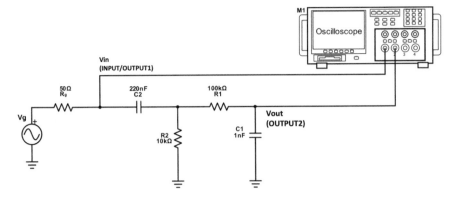

Fig. 2.19 Sketch of the experimental setup for the characterization of a CR-RC filter

- $R_2 = 10\,k\Omega$:
- $C_2 = 220\,nF$.

The transfer function of this filter is described by the following equation:

$$H(j\omega) = \frac{1}{1 + j\omega R_1 C_1} \frac{1}{1 + \frac{1}{j\omega R_2 C_2} + \frac{j\omega C_1}{j\omega C_2 - \omega^2 R_1 C_1 C_2}} .$$

It is the product of a high-pass and a low-pass transfer function, with a mixed term; but since $C_2 \gg C_1$, the mixed term can be neglected. So the transfer function is given by

$$H(j\omega) \simeq \frac{j\omega R_2 C_2}{1 + j\omega R_2 C_2} \frac{1}{1 + j\omega R_1 C_1} . \qquad (2.21)$$

The lower cutoff frequency is the one associated with the high-pass filter, so $f_{p,hp} \simeq 72\,Hz$; the upper cutoff frequency is the one related to the low-pass filter's behavior, so $f_{p,lp} \simeq 1.591\,kHz$. The bandwidth of the filter is simply the difference between the upper and lower cutoff frequencies:

$$B = f_{p,lp} - f_{p,hp} .$$

The general measurement procedure described in Sect. 2.3.1 has been applied. Table 2.7 summarizes all measurements taken. Amplitude uncertainties are reported in Table 2.8. Time and phase shift uncertainties are reported in Table 2.9.

It is possible to graphically compare the Bode diagrams of the measured data and the theoretical data, as shown in Fig. 2.20.

Table 2.7 Measurement data for the CR-RC filter

f (Hz)	V_i (V)	k_{V_i} (V/div)	V_o (V)	k_{V_o} (V/div)	Δt (μs)	k_t (μs/div)
5	10.30	2	0.643	0.1	−48000	10000
10	10.30	2	1.270	0.2	−23000	5000
25	10.30	2	3.06	0.5	−7800	1000
50	10.30	2	5.27	1	−3010.0	500
72	**10.30**	**2**	**6.51**	**1**	**−1646.0**	**200**
100	10.20	2	7.40	1	−908.0	200
200	10.20	2	8.70	2	−184.0	100
350	10.10	2	8.80	2	5.000	2
500	10.10	2	8.70	2	46.70	50
1000	9.10	2	7.00	1	72.80	10
1500	9.70	2	6.51	1	71.00	10
1600	**9.80**	**2**	**6.39**	**1**	**69.80**	**10**
2000	10.10	2	5.83	1	65.60	10
4000	10.20	2	3.58	0.5	45.80	10
10000	10.20	2	1.570	0.2	22.100	5
20000	10.20	2	0.780	0.2	11.600	2

Table 2.8 Evaluated uncertainty values for voltage amplitude measurements in a CR-RC filter, for each considered frequency value. The calculated values for A and the related uncertainties are also reported

f (Hz)	U_{Vi} (V)	$U_{r,Vi}$ (%)	U_{Vo} (V)	$U_{r,Vo}$ (%)	A_{dB} (dB)	$U_{A_{dB}}$ (dB)	A	U_A	$U_{r,A}$ (%)
5	0.56	5.4	0.028	4.4	−24.09	0.85	0.0624	0.0061	10
10	0.56	5.4	0.056	4.4	−18.18	0.86	0.123	0.012	10
25	0.56	5.4	0.14	4.6	−10.54	0.87	0.297	0.030	10
50	0.56	5.4	0.28	5.3	−5.82	0.93	0.512	0.055	11
72	0.56	5.4	0.28	4.3	−3.99	0.85	0.632	0.062	10
100	0.56	5.5	0.28	3.8	−2.79	0.81	0.725	0.067	9.3
200	0.56	5.5	0.56	6.4	−1.38	1.04	0.85	0.10	12
350	0.56	5.5	0.56	6.4	−1.20	1.03	0.87	0.10	12
500	0.56	5.5	0.56	6.4	−1.30	1.04	0.86	0.10	12
1000	0.56	6.2	0.28	4.0	−2.28	0.88	0.769	0.078	10
1500	0.56	5.8	0.28	4.3	−3.46	0.88	0.671	0.068	10
1600	0.56	5.7	0.28	4.4	−3.71	0.88	0.652	0.066	10
2000	0.56	5.5	0.28	4.8	−4.77	0.90	0.577	0.060	10
4000	0.56	5.5	0.14	3.9	−9.09	0.82	0.351	0.033	9.4
10000	0.56	5.5	0.056	3.6	−16.25	0.79	0.154	0.014	9.1
20000	0.56	5.5	0.056	7.2	−22.3	1.1	0.076	0.010	13

Table 2.9 Evaluated uncertainty values for time and phase shift measurements in a CR-RC filter, for each considered frequency value. The calculated values for $\Delta\phi$ and the related uncertainties are also reported

f (Hz)	$U_{\Delta t}$ (μs)	$U_{r,\Delta t}$ (%)	U_f (Hz)	$\Delta\phi$ (deg)	$U_{\Delta\phi}$ (deg)	$U_{r,\Delta\phi}$ (%)
5	161	0.34	0.00010	86.40	0.29	0.34
10	81	0.35	0.00020	82.80	0.29	0.35
25	16	0.21	0.00050	70.20	0.15	0.21
50	8.1	0.27	0.0010	54.18	0.15	0.27
72	3.2	0.20	0.0014	42.664	0.085	0.20
100	3.2	0.35	0.0020	32.69	0.12	0.36
200	1.6	0.87	0.0040	13.25	0.12	0.87
350	0.032	0.64	0.0070	−0.6300	0.0041	0.64
500	0.80	1.7	0.010	−8.41	0.14	1.7
1000	0.16	0.22	0.020	−26.21	0.059	0.22
1500	0.16	0.23	0.030	−38.34	0.088	0.23
1600	0.16	0.23	0.032	−40.205	0.094	0.23
2000	0.16	0.25	0.040	−47.23	0.12	0.25
4000	0.16	0.35	0.080	−65.95	0.23	0.35
10000	0.08	0.36	0.20	−79.56	0.29	0.37
20000	0.032	0.28	0.40	−83.52	0.23	0.28

2.3.5 Characterization of an LCR Filter

The LCR circuit, shown in Fig. 2.21, is a more selective band-pass filter, consisting of a cascade of a low-pass and a high-pass filter. The characteristic parameters of an LCR filter are the resonant frequency (f_0), the bandwidth (B), and the quality factor (Q), which are given by the following equations:

$$f_0 = \frac{1}{2\pi\sqrt{LC}}\,, \tag{2.22}$$

$$B = \frac{R}{2\pi L}\,, \tag{2.23}$$

$$Q = \frac{f_0}{B}\,. \tag{2.24}$$

The values of the electrical components used for implementing this LCR filter are:

- R = 10 Ω;
- C = 220 nF;
- L = 100 mH.

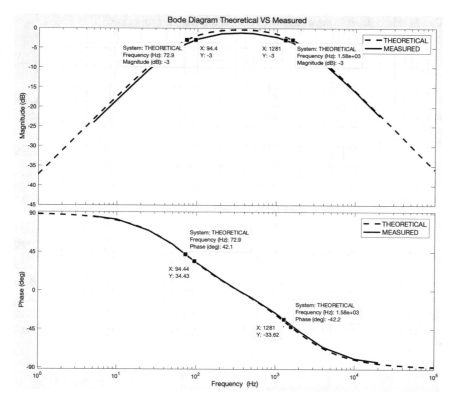

Fig. 2.20 Comparison between theoretical and measured Bode diagrams for the CR-RC filter

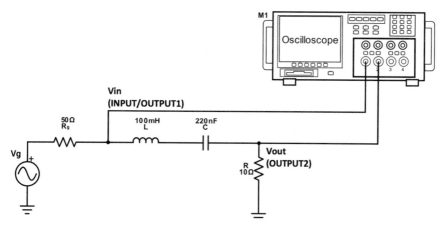

Fig. 2.21 Sketch of the experimental setup for the characterization of an LCR filter

Also in this case, it is very simple to obtain the transfer function:

$$H(j\omega) = \frac{j\omega RC}{(j\omega)^2 LC + j\omega RC + 1}. \tag{2.25}$$

The calculations of the theoretical values of resonance frequency, bandwidth, and quality factor are

$$f_0 = \frac{1}{2\pi \sqrt{LC}} = \frac{1}{2\pi \sqrt{0.1 \times 220 \times 10^{-9}}} \simeq 1073.02 \text{ Hz}, \tag{2.26}$$

$$B = \frac{R}{2\pi L} = \frac{10}{2\pi \times 0.1} \simeq 15.92 \text{ Hz}, \tag{2.27}$$

$$Q = \frac{f_0}{B} \simeq \frac{1073}{15.9} \simeq 67.4. \tag{2.28}$$

Again, the general measurement procedure described in Sect. 2.3.1 has been applied. Table 2.10 summarizes the measurement results. Amplitude uncertainties are reported in Table 2.11. Time and phase shift uncertainties are reported in Table 2.12.

It is possible to graphically compare the Bode diagrams of the measured data and the theoretical data, as shown in Fig. 2.22.

2.4 Didactic Laboratory Experiments on Operational Amplifiers

Similarly to the experiments for measuring the frequency-dependent responses of passive filters, in this section we address the practical use of DSO for characterizing some active devices such as operational amplifiers (theoretical details about operational amplifiers can be found in [4]). In particular, this section describes the experimental procedure adopted for measuring the most significant working parameters of operational amplifiers, such as voltage gain both in closed- and open-loop configurations, the Common-Mode Rejection Ratio (CMRR) and the slew rate. To this end, a general-purpose operational amplifier (i.e., μA 741 [5]) is considered, and its experimental response with respect to its ideal behavior is shown. Finally, as also carried out for the previous cases, the uncertainties affecting the experimental measurements are evaluated according to the theory described in Chap. 1, with the details given in Sect. 2.2. In these experiments, unlike those on passive filters, the power supply listed in Sect. 2.2 is used, but it does not give a meaningful contribution to the measurement uncertainties.

Table 2.10 Measurement data for the LCR filter

f (Hz)	V_i (V)	k_{V_i} (V/div)	V_o (mV)	k_{V_o} (mV/div)	Δt (μs)	k_t (μs/div)
100	10.11	2	13.93	2	−2350.0	500
500	9.92	2	82.6	20	−392.00	50
600	9.81	2	107.0	20	−292.50	50
700	9.64	2	135.6	20	−218.00	50
800	9.41	2	169	50	−148.40	50
900	9.13	2	199	50	−84.00	20
925	9.08	2	204	50	−68.00	10
950	9.03	2	209	50	−52.60	10
975	8.98	2	213	50	−37.80	10
1000	8.94	2	217	50	−24.500	5
1025	8.92	2	218	50	−11.100	5
1050	8.92	2	218	50	−1.000	1
1075	8.93	2	217	50	12.200	2
1100	8.95	2	216	50	25.150	5
1150	9.02	2	209	50	42.450	5
1200	9.11	2	201	50	57.00	10
1300	9.28	2	183	50	77.20	10
2000	9.85	2	96.0	20	92.00	10
5000	10.07	2	32.9	5	45.900	5
10000	10.08	2	16.2	5	24.200	5

2.4.1 Closed-Loop Gain

In this practical experience, the μA741 op-amp is used in its inverting configuration, as shown in Fig. 2.23. The electrical elements used for implementing this circuital configuration are:

- $R_1 = 3.3\,k\Omega$;
- $R_2 = 68\,k\Omega$;
- $R_3 = R_1//R_2 \simeq 3.147\,k\Omega$;
- $R_o = 10\,k\Omega$.

The first operation is offset reduction. To accomplish this, the input voltage V_i is set to zero. In this way, on the oscilloscope, only the output channel V_o is visualized, and due to the offset, the value is different from zero, despite the fact that $V_i = 0$. Therefore, connecting a potentiometer between the pin terminals devoted to the offset reduction (typically pin 1 and 5 on the IC) it is possible, through a fine adjustment, to minimize the offset voltage effect, which can be considered optimal when V_o is close to zero.

Table 2.11 Evaluated uncertainty values for voltage amplitude measurements in an LCR filter, for each considered frequency value. The calculated values for A and the related uncertainties are also reported

f (Hz)	U_{Vi} (V)	$U_{r,Vi}$ (%)	U_{Vo} (V)	$U_{r,Vo}$ (%)	A_{dB} (dB)	$U_{A_{dB}}$ (dB)	A	U_A	$U_{r,A}$ (%)
100	0.56	5.5	0.00072	5.2	-57.22	0.93	0.00138	0.00015	11
500	0.56	5.6	0.0072	8.7	-41.6	1.2	0.0083	0.0012	14
600	0.56	5.7	0.0072	6.7	-39.2	1.1	0.0109	0.0014	12
700	0.56	5.8	0.0072	5.3	-37.04	1.0	0.0141	0.0016	11
800	0.56	6.0	0.018	11	-34.9	1.4	0.0180	0.0030	17
900	0.56	6.1	0.018	9.0	-33.2	1.3	0.0218	0.0033	15
925	0.56	6.2	0.018	8.8	-33.0	1.3	0.0225	0.0034	15
950	0.56	6.2	0.018	8.6	-32.7	1.3	0.0231	0.0034	15
975	0.56	6.2	0.018	8.5	-32.5	1.3	0.0237	0.0035	15
1000	0.56	6.3	0.018	8.3	-32.3	1.3	0.0243	0.0035	15
1025	0.56	6.3	0.018	8.3	-32.2	1.3	0.0244	0.0036	15
1050	0.56	6.3	0.018	8.3	-32.2	1.3	0.0244	0.0036	15
1075	0.56	6.3	0.018	8.3	-32.3	1.3	0.0243	0.0035	15
1100	0.56	6.3	0.018	8.3	-32.3	1.3	0.0241	0.0035	15
1150	0.56	6.2	0.018	8.6	-32.7	1.3	0.0232	0.0034	15
1200	0.56	6.1	0.018	9.0	-33.1	1.3	0.0221	0.0033	15
1300	0.56	6.0	0.018	9.8	-34.1	1.4	0.0197	0.0031	16
2000	0.56	5.7	0.0072	7.5	-40.2	1.1	0.0097	0.0013	13
5000	0.56	5.6	0.0018	5.5	-49.7	1.0	0.00327	0.00036	11
10000	0.56	5.6	0.0018	11	-55.9	1.4	0.00161	0.00027	17

Due to the various types of noise present in the electrical components, an acceptable offset value is $<2\,\text{mV}$.

After compensating for offset, it is possible to evaluate the closed-loop gain of the amplifier.

The function generator is set to provide a sinusoidal signal of 1 kHz. The amplitude of the signal is changed until the output voltage on the oscilloscope is undistorted.

However, for a correct gain measurement, it is good practice to set the signal input (V_{ipp}) to a value that gives an output slightly below the nondistortion limit value. This is done to ensure that the operating condition is not in the saturation region of the amplifier.

2.4.1.1 Theoretical Evaluation of Closed-Loop Gain of μA741

For an op-amp in inverting configuration, the theoretical value of the gain (A_{CL}) is given by

Table 2.12 Evaluated uncertainty values for time and phase shift measurements in an LCR filter, for each considered frequency value. The calculated values for $\Delta\phi$ and the related uncertainties are also reported

f (Hz)	$U_{\Delta t}$ (μs)	$U_{r,\,\Delta t}$ (%)	U_f (Hz)	$\Delta\phi$ (deg)	$U_{\Delta\phi}$ (deg)	$U_{r,\Delta\phi}$ (%)
100	8.1	0.34	0.0020	84.60	0.29	0.34
500	0.81	0.21	0.010	70.56	0.15	0.21
600	0.81	0.28	0.012	63.18	0.18	0.28
700	0.81	0.37	0.014	54.94	0.20	0.37
800	0.80	0.54	0.016	42.74	0.23	0.54
900	0.32	0.38	0.018	27.22	0.10	0.39
925	0.16	0.24	0.019	22.644	0.054	0.24
950	0.16	0.31	0.019	17.989	0.056	0.31
975	0.16	0.43	0.020	13.268	0.057	0.43
1000	0.08	0.33	0.020	8.820	0.029	0.33
1025	0.08	0.72	0.021	4.096	0.030	0.73
1050	0.02	1.6	0.021	0.378	0.006	1.6
1075	0.03	0.27	0.022	−4.721	0.013	0.27
1100	0.08	0.32	0.022	−9.959	0.032	0.32
1150	0.08	0.19	0.023	−17.574	0.034	0.19
1200	0.16	0.28	0.024	−24.624	0.070	0.29
1300	0.16	0.21	0.026	−36.130	0.077	0.21
2000	0.16	0.18	0.040	−66.24	0.12	0.18
5000	0.08	0.18	0.10	−82.62	0.15	0.18
10000	0.08	0.33	0.20	−87.12	0.29	0.34

$$A_{CL} = -\frac{V_{opp}}{V_{ipp}} = -\frac{R_2}{R_1} . \qquad (2.29)$$

To evaluate the cutoff frequency (f_c), which corresponds to the passband B of the amplifier, it is important to consider the figure of merit GBP (Gain Bandwidth Product):

$$GBP = A_{CL} \times B .$$

For a μA741 amplifier, the GBP is approximately 1 MHz, so

$$f_c = B = \frac{GBP}{A_{CL}} = 45.543\,\text{kHz} . \qquad (2.30)$$

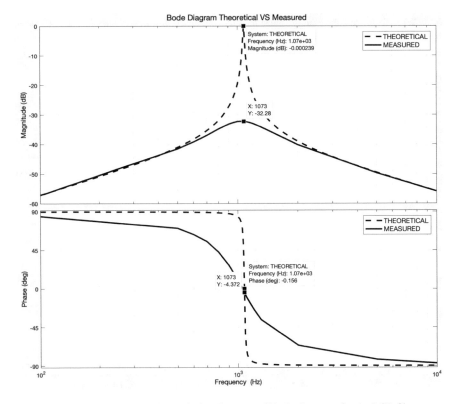

Fig. 2.22 Comparison between theoretical and measured Bode diagrams for the LCR filter

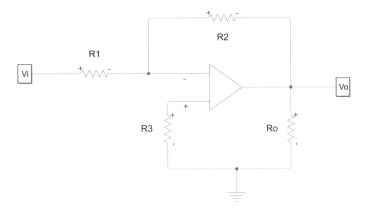

Fig. 2.23 Closed-loop configuration for offset reduction and gain measurement

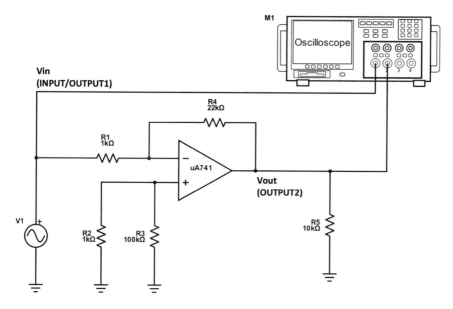

Fig. 2.24   Closed-loop configuration

2.4.1.2   Experimental Evaluation of Closed-Loop Gain of μA741

After these theoretical considerations, a practical experience is carried out using the circuit in Fig. 2.24.

The procedural steps are as follows:

1. Connect the function generator output to the "INPUT" port of the PCB, with a coaxial cable with BNC-BNC connectors.
2. Connect "OUTPUT1" and "OUTPUT2" of the PCB, respectively, to "CH1" and "CH2" of the oscilloscope, using BNC-BNC connectors. "CH1" of the oscilloscope will show the input signal of the amplifier, whereas "CH2" shows the output signal.
3. Manually set the attenuation factor of the probe of "CH1" and "CH2" to 1:1 (short BNC-BNC connectors are used instead of attenuation probes; therefore, the probe factor must be 1:1).
4. Reduce the offset of the amplifier to the minimum practically achievable, using the potentiometer as previously explained.
5. Select a sinusoidal signal on the function generator and set a starting frequency value (f).
6. Obtain the automatic measurement of the peak-to-peak voltage for both the "CH1" and "CH2" channels of the DSO.
7. Vary the frequency f of the sinusoidal signal on the function generator, and repeat step 6, to obtain several measurements, which is useful for characterizing the amplifier in a convenient frequency range.

Table 2.13 Measurement data for the closed-loop gain of the operational amplifier

f (Hz)	V_i (V)	k_{V_i} (V/div)	V_o (V)	k_{V_o} (V/div)
100	0.388	0.05	8.70	2
1000	0.388	0.05	8.70	2
5000	0.384	0.05	8.50	2
7000	0.382	0.05	8.40	2
9000	0.380	0.05	8.20	2
10000	0.380	0.05	8.10	2
15000	0.376	0.05	7.40	1
20000	0.376	0.05	6.67	1
25000	0.372	0.05	5.91	1
30000	0.372	0.05	5.19	1
50000	0.372	0.05	3.36	0.5
100000	0.370	0.05	1.75	0.5
700000	0.187	0.05	0.1550	0.02

Table 2.14 Evaluated uncertainty values for voltage amplitude measurements in closed-loop configuration for each considered frequency value. The calculated values for A and the related uncertainties are also reported

f (Hz)	U_{Vi} (V)	$U_{r,Vi}$ (%)	U_{Vo} (V)	$U_{r,Vo}$ (%)	A_{dB} (dB)	$U_{A_{dB}}$ (dB)	A	U_A	$U_{r,A}$ (%)
100	0.014	3.6	0.56	6.4	27.01	0.87	22.4	2.3	10
1000	0.014	3.6	0.56	6.4	27.01	0.87	22.4	2.3	10
5000	0.014	3.6	0.56	6.6	26.90	0.89	22.1	2.3	10
7000	0.014	3.7	0.56	6.7	26.84	0.90	22.0	2.3	10
9000	0.014	3.7	0.56	6.8	26.68	0.91	21.6	2.3	11
10000	0.014	3.7	0.56	6.9	26.57	0.92	21.3	2.3	11
15000	0.014	3.7	0.28	3.8	25.88	0.65	19.7	1.5	7.5
20000	0.014	3.7	0.28	4.2	24.98	0.69	17.7	1.4	7.9
25000	0.014	3.8	0.28	4.7	24.02	0.74	15.9	1.4	8.5
30000	0.014	3.8	0.28	5.4	22.89	0.80	14.0	1.3	9.2
50000	0.014	3.8	0.14	4.2	19.12	0.69	9.03	0.72	7.9
100000	0.014	3.8	0.14	8.0	13.5	1.0	4.73	0.56	12
700000	0.014	7.5	0.0056	3.6	−1.63	0.96	0.83	0.09	11

The obtained results are reported in Table 2.13.

The evaluated uncertainty values are summarized in Table 2.14.

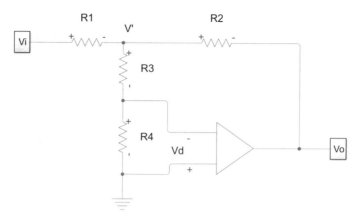

Fig. 2.25 Circuit scheme for open-loop gain measurements

2.4.2 Open-Loop Gain

The open-loop gain of an op-amp is defined as the ratio between the differential voltage applied between the inputs V_d and the output voltage V_o, in the absence of the feedback branch, i.e.,

$$A_{OL} = \frac{V_o}{V_d} .$$

The general procedural steps are identical to those for the measurement of the closed-loop gain. The important difference is that in open loop, it is impossible to achieve offset reduction, since even small variations of the adjustment potentiometer bring the output from positive saturation to negative and vice versa. In this experiment, an indirect measurement of V_d is used to measure the open-loop gain A_{OL}, using the schematic in Fig. 2.25.

The circuital components' values in the measurements are:

- $R_1 = R_3 = 15 \, \text{k}\Omega$;
- $R_2 = 100 \, \text{k}\Omega$;
- $R_4 = 15 \, \Omega$.

In practice, for the evaluation of the open-loop gain, the closed-loop configuration is still used, and the voltage divider, with $R_3 \gg R_4$, is exploited. The differential voltage is given by

$$V_d = V' \frac{R_4}{R_3 + R_4} .$$

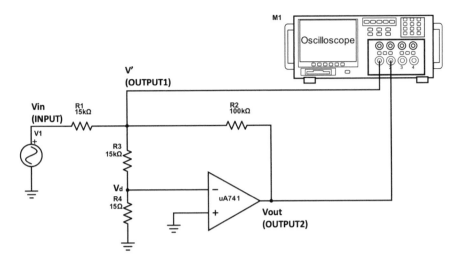

Fig. 2.26 Open-loop configuration

After V_d is computed, the open-loop gain is given by

$$A_{OL} = \frac{V_o}{V'} \frac{R_3 + R_4}{R_4} .$$

(2.31)

Thus, the obtained gain value can be considered coincident with the open-loop value.

2.4.2.1 Experimental Evaluation of Open-Loop Gain of μA741

For this experience, the circuit shown in Fig. 2.26 is considered. Tables 2.15 and 2.16 summarize the measurement results and uncertainty evaluation results, respectively.

2.4.3 Common-Mode Rejection Ratio

The common-Mode Rejection Ratio (CMRR) of an amplifier measures the device's tendency to reject input signals common to both inputs. A high CMRR is important in applications in which the signal of interest is represented by a small fluctuation of the voltage around a voltage offset (possibly high) or when the relevant information is contained in the difference between two voltage signals.

The CMRR is expressed as the ratio between the differential mode gain (A_d) and the common-mode gain (A_{cm}), i.e.,

Table 2.15 Measurement data for the open-loop gain of the operational amplifier

f (Hz)	V' (V)	$K_{V'}$ (V/div)	V_o (V)	k_{V_o} (V/div)
3	0.0258	0.005	5.47	1
5	0.0390	0.01	5.37	1
8	0.0600	0.01	5.33	1
10	0.0738	0.01	5.31	1
25	0.165	0.05	4.84	1
50	0.260	0.05	3.87	0.5
75	0.309	0.05	3.04	0.5
100	0.332	0.05	2.47	0.5
250	0.365	0.05	1.080	0.2
500	0.372	0.05	0.557	0.1
750	0.372	0.05	0.367	0.05
1000	0.372	0.05	0.275	0.05
5000	9.300	0.2	1.350	0.2
50000	9.070	2	0.137	0.05
190000	7.710	1	0.0394	0.01

Table 2.16 Evaluated uncertainty values for voltage amplitude measurements in open-loop configuration for each considered frequency value. The calculated values for A and the related uncertainties are also reported

f (Hz)	$U_{V'}$ (mV)	$U_{r, V'}$ (%)	V_d (mV)	U_{V_d} (mV)	U_{r, V_d} (%)	U_{V_o} (V)	U_{r, V_o} (%)	A_{dB} (dB)	$U_{A_{dB}}$ (dB)	$U_{r, A}$ (%)
3	1.8	7.0	0.0258	0.0044	17	0.28	5.1	106.5	1.1	12
5	3.6	9.2	0.0390	0.0075	19	0.28	5.2	102.8	1.3	14
8	3.6	6.0	0.0599	0.0096	16	0.28	5.3	99.0	1.0	11
10	3.6	4.9	0.074	0.011	15	0.28	5.3	97.15	0.88	10
25	18	11	0.165	0.034	21	0.28	5.8	89.4	1.5	17
50	18	6.9	0.260	0.044	17	0.14	3.6	83.46	0.92	11
75	18	5.8	0.309	0.049	16	0.14	4.6	79.87	0.91	10
100	18	5.4	0.332	0.051	15	0.14	5.7	77.4	1.0	11
250	18	4.9	0.365	0.054	15	0.056	5.2	69.43	0.88	10
500	18	4.8	0.372	0.055	15	0.028	5.0	63.51	0.86	10
750	18	4.8	0.372	0.055	15	0.014	3.8	59.89	0.75	8.7
1000	18	4.8	0.372	0.055	15	0.014	5.1	57.38	0.86	10
5000	72	0.77	9.3	1.0	11	0.056	4.1	43.25	0.43	4.9
50000	720	7.9	9.1	1.6	18	0.014	10	23.6	1.6	18
190000	360	4.7	7.7	1.1	15	0.0028	7.1	14.2	1.0	12

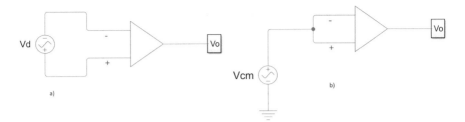

Fig. 2.27 **a** Differential signal and **b** common-mode signal

$$\text{CMRR} = \frac{A_d}{A_{cm}} , \qquad\qquad (2.32)$$

where:

- $A_d = V_o/V_d$, namely the ratio between the output voltage and the difference between the signal voltages applied to the two inputs;
- $A_{cm} = V_o/V_{cm}$, namely the ratio between the output voltage and the simultaneous voltage applied to the two inputs.

Figure 2.27 illustrates the different schematizations for the input of a differential or common-mode signal.

With such signals, it is therefore possible to obtain the differential and common-mode gains using the circuital schemes shown in Fig. 2.28.

It is evident that in this case, the differential gain A_d corresponds to the closed-loop gain A_{CL} (as in (2.29)). For the common-mode gain A_{cm} evaluation, however, the condition $R_2/R_1 = R_4/R_3$ must be satisfied within a certain tolerance limit.

The CMRR value can also be expressed in decibels, through the relationship

$$\text{CMRR}_{dB} = 20 \log_{10}\left(\frac{A_d}{A_{cm}}\right).$$

Theoretically, the CMRR should be ∞, since A_{cm} should be null; in practice, the value of the CMRR is generally in the range 60–120 dB.

2.4.3.1 Experimental Evaluation of CMRR

The laboratory test is performed using the PCB shown in Fig. A.9 of Appendix A. Figure 2.29 shows the circuit diagram used for the measurement of the CMRR. The input signal is a sinusoidal voltage at 1 kHz.

For this type of measurement, the initial procedures are the same as the previous ones. However, in this case, they concern the evaluation of the gains of both possible configurations of the circuit changing the switch position. The results are summarized in Tables 2.17, 2.18, and 2.19.

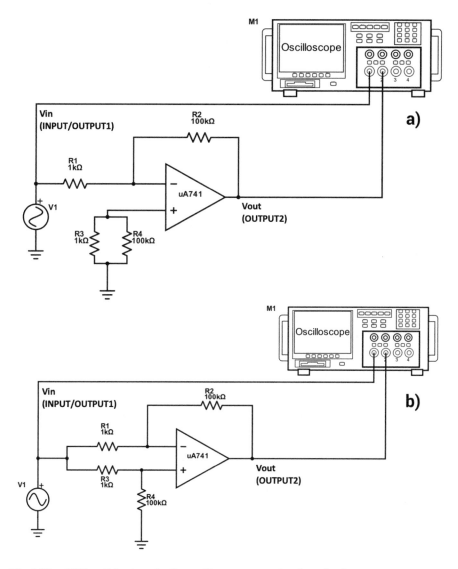

Fig. 2.28 a Differential gain evaluation and **b** common-mode gain evaluation

2.4.4 Slew Rate

The Slew Rate (SR) of an op-amp is defined as the time taken by the amplifier to make a variation of the output voltage. It is expressed as

$$SR = \frac{\Delta V_o}{\Delta t}.$$ (2.33)

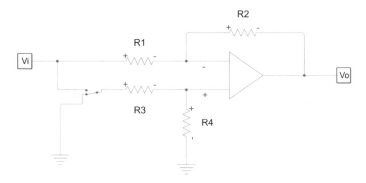

Fig. 2.29 Circuit diagram used for the evaluation of the CMRR

Table 2.17 Measurement data for the CMRR of the operational amplifier

Mode	f	V_i	k_{V_i}	V_o	k_{V_o}
–	(Hz)	(V)	(V/div)	(V)	(V/div)
Differential	1000	0.0964	0.02	9.60	2
Common	1000	6.05	1	0.01060	0.002

Table 2.18 Evaluated uncertainty values for voltage amplitude measurements in differential and common modes of the operational amplifier

f (Hz)	U_{V_i} (V)	U_{r, V_i} (%)	U_{V_o} (V)	U_{r, V_o} (%)	A_{dB} (dB)	$U_{A, dB}$ (dB)	A	U_A	$U_{r, A}$ (%)
1000	0.0056	5.8	0.56	7.5	40.0	1.0	99.59	13.25	13
1000	0.28	4.6	0.00072	6.8	−55.1	1.0	0.00175	0.00020	11

Table 2.19 CMRR measurement results and related uncertainty values

CMRR	U_{CMRR}	$U_{r,CMRR}$ (%)	$CMRR_{dB}$ (dB)	$U_{CMRR,dB}$ (dB)
57E+3	13E+3	23	95.1	2.0

The output switching speed established by the slew rate is the maximum possible for an operational amplifier: if the input signal has a trend that forces the output to a higher switching speed, it is distorted. As a result, a sinusoidal input can become a triangular wave, for example.

It is also possible to demonstrate that if the slew rate and the maximum variation of the output voltage $\Delta V_{o_{max}}$ are known, then the maximum frequency f_{max} of the sinusoidal signal that leads to undistorted output is

$$f_{max} = \frac{SR}{2\pi \, \Delta V_{o_{max}}} \, . \tag{2.34}$$

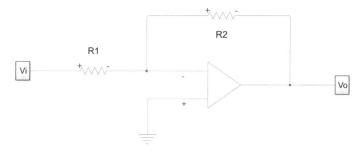

Fig. 2.30   Circuit diagram used for the evaluation of the slew rate

Table 2.20   Measurement data for the slew rate of the operational amplifier

f (Hz)	ΔV (V)	k_V (V/div)	Δt (μs)	k_t (μs/div)
1000	12.4	5	23.76	20

Table 2.21   Evaluated uncertainty values for the slew rate. Voltage amplitude and time uncertainty values are also reported

$U_{\Delta V}$ (V)	$U_{r,\Delta V}$ (%)	$U_{\Delta t}$ (μs)	$U_{r,\Delta t}$ (%)	SR (V/μs)	U_{SR} (V/μs)	$U_{r,SR}$ (%)
1.4	11	0.32	1.3	0.522	0.066	13

Typical slew rate values range from a few tens to a few hundreds of volts per microsecond. The slew rate is evaluated in the closed-loop configuration shown in Fig. 2.30.

2.4.4.1   Measurement of the Slew Rate

For this measurement, the same configuration of the closed loop gain (Fig. 2.24) must be used.

The procedural steps in this experience are almost identical to the previous ones. The difference in this case is that the signal provided by the function generator is a square wave.

To quantify the slew rate, a double-cursor measurement must be performed on the rise time and the descent time of the output signal.

These measurements are related to the transition times. This means that they are performed based on specific threshold voltage levels. The oscilloscope threshold voltage levels for these measurements are 10% and 90% of the V_{min} and V_{max} values of the square wave, respectively. The measurement results and the related uncertainty values are summarized in Tables 2.20 and 2.21, respectively.

References

1. Laughton MA, Warne DF Electrical engineer's reference book, 16th edn. Newnes, Oxford. Accessed 27 Sept 2002
2. Huijsing JH, van der Plassche RJ, Sansen W Analog Circuit Design. Springer, Berlin
3. InfiniiVision 2000 X-Series Oscilloscopes datasheet. Keysight Technologies, USA. https://www.upc.edu/sct/es/documents_equipament/d_332_dsox2012a.pdf. Accessed 2 Aug 2014
4. Huijsing J (2001) Operational amplifiers theory and design. Springer, Berlin
5. μA741 general-purpose operational amplifiers. http://www.ti.com/lit/ds/slos094g/slos094g.pdf
6. Programmable DC Power Supplies datasheet. Keysight E363xA Series, USA. http://literature.cdn.keysight.com/litweb/pdf/5968-9726EN.pdf. Accessed 8 May 2018

Chapter 3
Frequency Domain Measurements

Abstract Theoretically, the transformation from the time domain (TD) to the frequency domain (FD) does not involve any loss of information of the signal, but only a different representation of it. Depending on the application, the analysis of a signal in FD may be more useful for the characterization of signals and devices, since it separates the different components of the signal. The instruments that are typically used for measurements in FD are spectrum analyzers (SAs) or, alternatively, the fast Fourier transform (FFT) algorithm of digitized time-domain signals. This chapter begins with a brief overview of the theoretical background on spectral analysis and on the measurement instruments operating in the frequency domain. Then several educational laboratory experiments are described that use spectral analysis with analog and digital approaches for characterizing signals, for assessing aliasing, for quantifying the harmonic distortion and the signal-to-noise ratio, for measurements on modulated signals, etc.

3.1 Basic Theory and Functionalities of Spectrum Analyzers

The spectrum analyzer (SA) is a fundamental instrument for the analysis of signals in the FD. This device, which can be either analog or digital, has different operating techniques. With an SA it is possible to display the harmonic content of a signal, to characterize various types of electronic circuits, to measure signal-to-noise ratios, to check the emission levels of electromagnetic (EM) disturbances of electronic devices, etc.

TD and FD represent two different perspectives of the same "object". Figure 3.1 shows the dual representation in the time domain and in the frequency domain of a signal that includes two sinusoidal components.

A spectrum analyzer allows the frequency representation of a signal in two different ways:

© Springer Nature Switzerland AG 2020   81
A. Cataldo et al., *Basic Theory and Laboratory Experiments in Measurement and Instrumentation*, Lecture Notes in Electrical Engineering 663,
https://doi.org/10.1007/978-3-030-46740-1_3

Fig. 3.1 Dual representation
of a signal in TD and FD

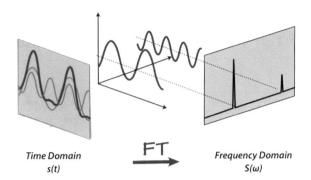

Time Domain   FT   Frequency Domain
s(t)   S(ω)

- Swept-tuned analyzers: In this case, the super-heterodyne technique is the most
  commonly implemented. This method allows one to obtain a frequency-translated
  version of the harmonic components of the signal in a different band from the
  starting one. After the frequency translation, the signal is filtered by a band-pass
  filter having a frequency band much smaller than that of the input signal. This
  technique is used for implementing spectral analysis through an analog modality.
- Fast Fourier transform (FFT) (real-time analyzers): This second method digitizes
  the signal in the time domain and displays it in the frequency domain, using the
  Discrete Fourier Transform (DFT) for a finite number of samples. The practical
  result of this operation is a simultaneous parallel filtering on the whole useful
  frequency band. This kind of technique is at the basis of digital spectral analysis.
  Fourier analysis produces the decomposition of the overall waveform into distinct
  spectral components, each with specific values of frequency, amplitude, and phase.

The advantages of the second technique are:

1. significant increase in speed (the instruments that implement this technique are
   called "real time" analyzers);
2. the possibility of analyzing single-shot signals (a consequence of the previous
   point);
3. the possibility of measuring phase and amplitude of each individual harmonic.

On the other hand, the disadvantages compared to swept-tuned analyzers are:

1. a lower frequency band (compared to a swept-tuned analyzer with similar cost);
2. reduced sensitivity and dynamic range values.

3.1.1 Analog Spectrum Analyzers

Considering the duality of TD and FD, it immediately follows that the simplest and
most immediate way to perform the analysis of the harmonic content of a signal
would be to construct a band-pass filter. This filter would have to be very selective,

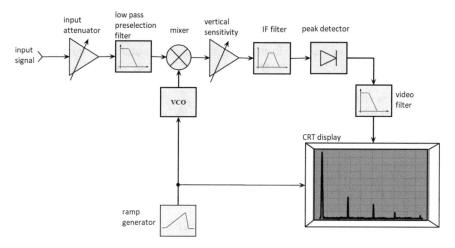

Fig. 3.2 Scheme of the super-heterodyne analyzer

but with a variable and electronically-controllable central frequency. An essential characteristic of a band-pass filter is the *selectivity* or *quality factor*. It can be defined as the ratio of the frequency bandwidth at a defined attenuation (B_2) and the width of its pass band (B_1):

$$\text{Selectivity} \doteq Q = \frac{B_2}{B_1}. \tag{3.1}$$

As a general figure of merit it can be assumed that the filter performance is better when Q is close to 1.

The most common analog spectrum analyzer is the so-called *super-heterodyne analyzer*. Figure 3.2 represents the basic functional scheme of a super-heterodyne analyzer.

The functional blocks are:

- *Input attenuator*: It attenuates the amplitude of the input signal to the sensitivity of the successive blocks. Thanks to this block, the dynamic range of the SA can be several tens of dB. The attenuation of the input signal allows one to reduce the distortion that is introduced by the mixer for significant power signals.
- *Low-pass preselection filter*: This block filters the input signal to eliminate signals outside the SA's band.
- *VCO (Voltage Controlled Oscillator)*: This generates a sinusoidal (or cosinusoidal) signal, whose frequency is varied following a linear ramp.
- *Mixer*: The mixer is the "core" of the instrument, and it allows the frequency translation of the input signal so that it can be analyzed by a fixed selective band-pass filter. Thanks to the mixer, it is possible to overcome the limit due to the selective band-pass filters with central frequency variable in wide ranges.
- *Ramp generator*: This generates the linear ramp that is used to sweep the VCO linearly from a minimum frequency ($f_{LO_{min}}$) up to a maximum frequency ($f_{LO_{max}}$).

- *Vertical sensitivity*: This is an adjustable gain amplifier. By default, it usually operates "in sync" with the input attenuator (this means that it amplifies as much as the attenuator attenuates. In this way, any distortion effects introduced by the mixer are not amplified. However, the operator is free to set a different value of the gain.
- *IF (intermediate frequency) filter*: This is a band-pass filter with a fixed central frequency (f_{IF}) but with a bandwidth that can be controlled by the operator. The bandwidth of the IF filter is called the *Resolution Bandwidth* (RBW); the RBW is controlled automatically by the instrument or by the operator.
- *Peak detector*: This block allows to measure the amplitude of the spectral components, if the IF filter is well designed and if it is sufficiently selective, this stage receives an essentially sinusoidal signal and, for a given VCO frequency, returns a signal with very low residual ripple.
- *Video filter*: This is a low-pass filter that reduces the noise introduced by the previous blocks of the SA. This improves the quality of the displayed signal. This filter strongly attenuates the noise superimposed on the signal in the various intermediate stages of the analyzer.
- *Display*: This serves to visualize the result of the spectral analysis.

The frequency mixing or multiplication is the key feature of the super-heterodyne analyzer. In fact, the local oscillator enables signals to be translated in frequency, thereby enabling signals to be converted up or down in frequency. The original frequency content of the input signal is translated to another frequency, so as to be sequentially analyzed by the frequency filter. The frequency of the local oscillator governs the frequency of the signal that will pass through the intermediate frequency filter. This is swept in frequency (linearly increasing in frequency) so that it covers the required band. The sweep voltage used to control the frequency of the local oscillator also controls the sweep of the scan on the display. In this way, the position of the scanned point on the screen relates to the position or frequency of the local oscillator and hence the frequency of the incoming signal. As a result of the frequency convolution that occurs in the mixer between the input signal $X(f)$ and the frequency-variable sinusoidal signal generated by the VCO, the original spectrum is shifted, on the frequency axis, by a quantity equal to the f_{LO} frequency of the signal supplied by the VCO. In this way, the spectrum of the signal at the mixer output is equal to $Y(f) = X(f \pm f_{LO})$ ($\pm f_m$ are the limit frequencies of the spectrum), as shown in Fig. 3.3.

The frequency-shifted signal, after a possible amplification, is then sent sequentially to the highly selective IF filter, which in practice, analyzes a very small frequency band portion. If f_s is a generic frequency component of the signal contained in the frequency bandwidth of the SA ranging from $f_{s_{min}}$ to $f_{s_{max}}$, the output of the mixer will contain the frequencies ($f_{LO} \pm f_s$). Generally, in most SAs, the value of f_{IF} is lower than the minimum frequency $f_{LO_{min}}$ generated by the local oscillator. The only components of the input signal that can pass through the IF filter are those with a frequency f_s such that the difference $f_{LO} - f_s$ falls within the pass-band of the filter itself. When the frequency f_{LO} varies (swept by the linear ramp), this

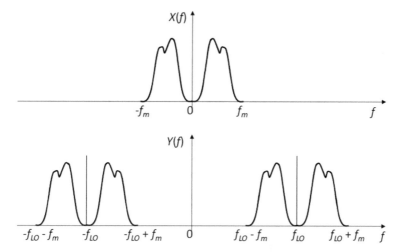

Fig. 3.3 Comparison between the original input signal spectrum and that obtained by the frequency mixing with the VCO signal

condition is reached, sequentially, for the different spectral components of the input signal.

In other words, given f_s, we have

$$f_s = f_{LO} - f_{IF} , \tag{3.2}$$

and in particular,

$$f_{s_{min}} = f_{LO_{min}} - f_{IF} \qquad\qquad f_{s_{max}} = f_{LO_{max}} - f_{IF} , \tag{3.3}$$

where $f_{LO_{min}}$ is the frequency value that translates the lowest frequency component of the signal spectrum ($f_{s_{min}}$) to f_{IF}; similarly, $f_{LO_{max}}$ is the frequency value that translates to f_{IF} the component with the highest upper frequency of the signal spectrum.

Figure 3.4 shows a schematization of the working principle of the swept-spectrum analysis. It is important to note that as a direct consequence of the working principle, the 0 Hz component cannot be analyzed, since the lower frequency of the instrument is intrinsically limited to a value of $f_{s_{min}}$ of approximately only a few kilohertz.

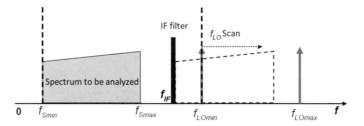

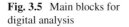

Fig. 3.4 Sketch of the operating principle of the swept-spectrum analysis

Fig. 3.5 Main blocks for
digital analysis

3.1.2 Digital Spectrum Analysis

The general scheme for a spectrum analysis implemented digitally can be associated, in practice, with the basic scheme of the digital oscilloscope. As a matter of fact, if the postprocessing block includes the possibility to execute the FFT algorithm, the oscilloscope can work also in the frequency domain as a digital spectrum analyzer. The main blocks are recalled in Fig. 3.5:

- *Transduction*: This can be included if necessary at the probing level.
- *Preprocessing*: This consists of electronic amplifiers, anti-aliasing filters, low-pass filters (for eliminating noise), and high-pass filters (for eliminating the DC component).
- *Conversion*: This is an ADC "fast" or "very fast" (some GSa/s) with low resolutions (typical value: 8 bits).
- *Postprocessing*: Window functions (Hanning, Hamming, Kaiser, Flat-Top windows, etc.) are applied and the fast Fourier transform (FFT) is calculated. Furthermore, this section performs calculations on the spectrum (effective value of the single harmonics, effective value of the noise "carpet", THD, etc.).
- *Display*: This displays the result of the spectral analysis.

Clearly, the most important section of the instrument is the postprocessing block. The discrete Fourier transform (DFT) establishes the relationship between the samples of the signal in the time domain and their representation in the frequency domain. Therefore, if there are N samples in the time domain, the DFT will be composed of N samples. Therefore, the spectrum of the sampled signal is given by

$$X(mF) = \sum_{n=0}^{N-1} x(nT_c) \cdot e^{-j2\pi nmFT_c} , \qquad (3.4)$$

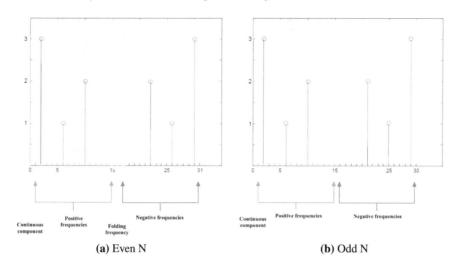

Fig. 3.6 Comparison between a spectral response of a periodic signal discretized with an even number of samples N (**a**) and an odd number of samples N (**b**)

where T_c is the sampling period and F is the frequency.

The distance between two successive samples over time is equal to the sampling period T_c, while the distance between two samples in the frequency domain is equal to the spectral resolution ΔF.

Since the digital spectrum analyzer processes only real signals over time, its DFT is complex and contains both amplitude and phase information:

$$X(mF) = |X(mF)| \, e^{j\angle X(mF)} . \tag{3.5}$$

Moreover, since the signal is real, its DFT is symmetric with respect to the $N/2$ index:

$$|X[mF]| = |X[(N-m)F]| \angle X[mF] = -\angle X[(N-m)F] . \tag{3.6}$$

Therefore, the DFT of a real signal contains redundant information in its N samples: the use of $\frac{N}{2}$ samples is sufficient for a complete representation of the signal in the FD.

If N is even, the useful spectral components are $\frac{N}{2} - 1$. The component at $\frac{N}{2}$ represents the folding frequency, while the components from $\frac{N}{2} + 1$ to $N - 1$ are the mirrored components. If N is an odd number, the same considerations hold for the interpretation of spectral lines; the only difference is that the folding component is invisible, because $\frac{N}{2}$ is not an integer.

The difference between N even and N odd is shown in Fig. 3.6.

It is worth underlining that to obtain a considerable savings of time in digital processing, the implementation of the DFT is done effectively through the FFT algorithm. This algorithm is effective when N is a power of 2, and it exploits all the

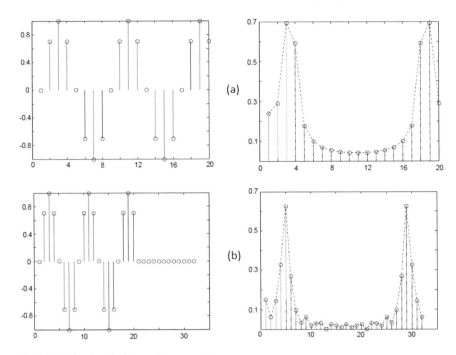

Fig. 3.7 With (**a**) and without (**b**) zero-padding

symmetries of the DFT. If N is not a power of 2, various techniques can be used to allow the use of the FFT; the most widespread artifice is *zero-padding*, i.e., the addition of zero-value samples at the end of the sequence, speeding up the processing and "fictitiously" increasing the resolution. This operation actually produces only a frequency interpolation. The technique of zero-padding is illustrated in Fig. 3.7.

The DFT provides all information on the amplitudes and phases of the individual harmonic components of the signal. In all those cases in which the signal phase is not important, it may be convenient to consider the *Power Spectral Density (PSD)*. The PSD is obtained by squaring the individual components of the DFT and displaying the power of the individual harmonic components with the frequency.

As theoretically predicted by the well-known sampling theorem, ideally perfect reconstructibility is possible for:

- an aperiodic signal, considering an infinite number of samples;
- a periodic band-limited signal having a finite number of samples that are contained precisely in a window whose time duration is an integer multiple of the signal period (i.e., coherent sampling condition).

The sample number of the DFT must be a finite number N, equivalent to the number of samples in the time domain. This "finite limitation" involves an inherent restriction of the observation interval with a "periodicization" of the signal with period N. This assumption, which is applied in the cases of both periodic signal and aperiodic

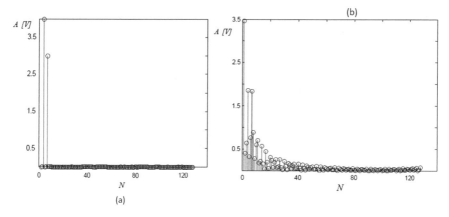

Fig. 3.8 a Original signal spectrum and **b** frequency-spread spectrum of windowed signal

signal, generally produces a discontinuity between some successive periods (real or fictitious) of the acquired signal. The result of this operation in the frequency domain is the appearance of spectral components that actually are not present in the spectrum of the original signal. This unwanted effect is a direct consequence of the time-windowing of the signal, or equivalently, of the corresponding frequency convolution between the original spectrum of the signal and that of the window signal. In this way, the final spectrum is a "spread" version of the original one, as represented in Fig. 3.8.

As a consequence of the aforementioned effect, two possible errors can be introduced:

- a *truncation error*, with reference to the phenomenon in the time domain, or *leakage* (spectral dispersion), with reference to the phenomenon in the frequency domain;
- a possible *aliasing error*, due to the folding of the "added" spectral components caused by truncation.

To reduce this phenomenon, it is necessary to use appropriate weight functions, otherwise called *windows*. The task of windows is to attenuate the discontinuities introduced by a time-limited observation; this result is obtained using weight functions that present a "gentler" trend at the edges of the observation interval. The windowing implies the multiplication over time of the stored signal samples with the weight function. Consequently, it corresponds to the convolution between the signal spectrum and the window spectrum in the frequency domain. Only when the data record contains an integer number of periods, regardless of the window used, are the effects of the leakage not present and the spectral components correctly positioned in the frequency domain (and not spread throughout it).

It is important to understand that there is no "ideal" window for all applications, because each window makes a compromise between two factors:

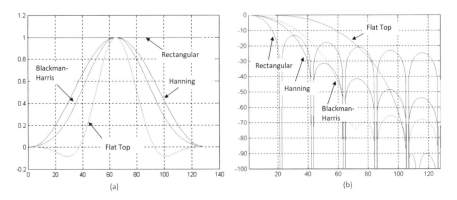

Fig. 3.9 Windows in **a** time domain and **b** frequency domain

- *frequency resolution* (width of the main lobe);
- *spectral dispersion* (amplitude and decay of the side lobes).

The width of the main lobe and the amplitudes of the side lobes cannot be simultaneously optimized for the same duration of time.

The main windows available on a digital spectrum analyzer are *Hanning*, *Flat Top*, *Rectangular*, and *Blackman–Harris* (Fig. 3.9).

An additional note refers to the typical settings of an SA, which include:

- *Horizontal and vertical axes.* The horizontal axis is directly calibrated in Hertz, and it represents the frequency. The vertical axis is calibrated in volts or watts, and there is the possibility to use a linear or logarithmic scale [dB]. For signals that differ by a few tens of dB, the linear scale can be used, while for signals that differ by many tens of dB, it is necessary to use the logarithmic scale (calibrated in dBV, dBmV, or dBμV).
- *Frequency range.* This is the extension of the frequency band in which the instrument is able to perform measurements with the specified uncertainty. For superheterodyne analog spectrum analyzers, the frequency range typically varies from a few kHz (lower band limit $f_{s_{min}}$) up to several tens of GHz (upper limit to $f_{s_{max}}$). Digital spectrum analyzers can cover a frequency bandwidth ranging from 0 Hz to several GHz. The so-called hybrid spectrum analyzers are another commercially available instrumental option, and they include both the advantages of the analog analyzer (wide bands) and those of the digital analyzer (high resolutions).
- *Resolution.* This refers to the minimum frequency separation that two spectral components must have to be correctly displayed by the SA. For the *analog spectrum analyzer*, the resolution is mostly determined by the "shape" of the frequency response of the IF filter, i.e., its bandwidth (RBW) and its selectivity. Frequency resolution also, however, depends on other aspects. For example, if there is a small amplitude difference between two close harmonics, the smaller harmonic can literally disappear under the side tails of the larger harmonic. It is necessary to keep in mind that the IF filter is a dynamic device; its output, at a defined instant

t_0, depends on the states that the input and output have assumed for $t < t_0$. This implies that the output of the IF filter is brought up to speed only after a certain time from the application of a fixed input; this "memory" of the IF filter, called settling time, is to a first approximation, inversely proportional to the RBW:

$$T_s \text{ (ns)} \approx \frac{0.35}{RBW \text{ (MHz)}} . \tag{3.7}$$

The smaller the RBW used (hence the better the frequency resolution), the more time it will take to complete the analysis over a fixed frequency span.

For the *digital spectrum analyzer*, the frequency resolution coincides with the frequency resolution of the FFT implemented by it:

$$\Delta f = \frac{f_c}{N} = \frac{1}{N \cdot T_c} . \tag{3.8}$$

The resolution can be increased either by reducing the sampling frequency, f_c, or by increasing the number of samples, N, to be processed. Both of these operations can be manually adjusted on the analyzer, but they involve two possible problems:

1. an increase in the number of samples N slows down the FFT processing;
2. a reduction in the sampling frequency may result in the appearance of aliasing.

- *Distortion.* This is an index of the alteration introduced by the instrument on the spectrum of the signal under examination. It is generated by the nonlinearities of the circuits that the signal passes through. For the analog spectrum analyzer, the most critical stage is the mixer, while in the digital spectrum analyzer, the nonlinearities are introduced by the analog preconditioning stages and by the ADC. The most common way of giving distortion specifications is to quantify the amplitudes of the second and third spurious harmonics introduced into the spectrum. Distortion is normally given in dBc (decibels relative to carrier, i.e., referring to the fundamental).
- *Dynamic range.* This is the difference between the maximum and minimum signal amplitudes that the instrument is able to detect in the frequency domain; the dynamic range is limited below by the total noise generated inside the instrument and above by its distortion. Normally, it is given in dB. For the digital spectrum analyzer, the dynamic range is limited by the characteristics of the A/D converter, to which are added the negative effects of noise and nonlinearity of the oscilloscope (all-in-one instrument). Any technique that is able to reduce the background noise allows one to increase the dynamic range; so, for example, an increase in the number of samples for the FFT (zero padding) reduces the noise floor (the total noise power is spread over a greater number of FFT outputs).

- *Uncertainty* (both in frequency and in amplitude). Just as for an oscilloscope, the measurement uncertainty is specified for time and amplitude measurements. In both cases, the specifications are given both for absolute measures and for relative measures. The first are the direct ones, executed with a single cursor and are, for example, the power measurements of an RF carrier, its frequency measurements, etc. The latter are the indirect ones, performed using two cursors simultaneously, for example distortion or distance measurements of harmonics or subharmonics from the carrier. Also in this case, a careful analysis of the various uncertainty components can lead to a reduction in the overall uncertainty of the instrument.
- *Sensitivity.* This is a measure of the instrument's ability to detect small signals. It is usually specified in dBm. The sensitivity is mostly limited by the noise generated by the analyzer's internal circuitry.

3.2 FFT-Based Measurements

This section addresses the digital spectral analysis carried out through FFT-based processing of the TD signals acquired through the same oscilloscope previously introduced in Chap. 2. To this end, the dedicated software (internal to the oscilloscope) was used.

In particular, the first experience addresses the aliasing effect in periodic signals and the appearance of "spurious" components corresponding to particular frequency values.

As briefly described in the previous section, for the calculation of the FFT, it is possible to consider different types of available window functions. This possibility is also available in the oscilloscope's software. Hence the following experience compares the results obtained from applying different window functions on the same TD signal for the transformation.

3.2.1 FFT on Periodic Signals and Aliasing Effect

Equipment:

1. a digital oscilloscope with FFT functionality (Keysight InfiniiVision DSO X-2012A)
2. a function generator (HP 33120A) [1]
3. connecting cables and adapters as necessary.

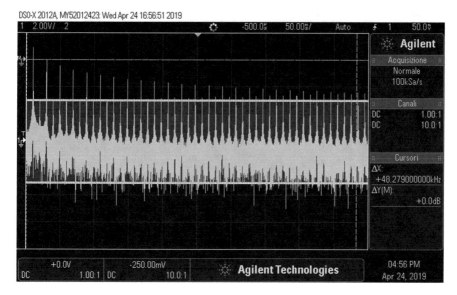

Fig. 3.10 FFT of a square wave at $f_0 = 990\,\text{Hz}$, as displayed by the oscilloscope

Tasks:

Use the function generator to generate a square wave with frequency $f_0 = 990\,\text{Hz}$ and visualize it with the oscilloscope; then:

1. Calculate and display its FFT through the dedicated software available in the oscilloscope;
2. Analyze and comment on the phenomenon of aliasing as it appears in the case of step 1;
3. Select different window functions from the oscilloscope menu (e.g., rectangular, flat top, Blackman–Harris, etc.) and compare the results of the FFT processing.

The procedural steps are as follows:

Generate a square waveform with frequency $f_0 = 990\,\text{Hz}$ and set the sampling frequency of the oscilloscope to $f_c = 100\,\text{kHz}$ (100 kSa/s).

The spectrum of a square wave is characterized by the presence of the ***harmonics*** at odd multiples of the fundamental frequency (f_0; 3 f_0; 5 f_0; ...), as shown in Fig. 3.10. Immediately before the folding frequency ($f_c/2 = 50$ kHz), the following harmonics will certainly be present:

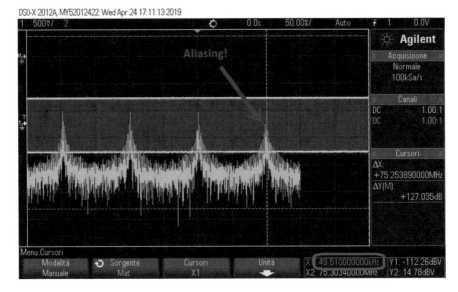

Fig. 3.11 FFT of the square wave signal, highlighting the first spurious harmonic

- the harmonic component corresponding to $49 \cdot f_0 = 48510\,\text{Hz}$;
- the 50th harmonic (even) will be at $49500\,\text{Hz}$, but strongly attenuated;
- the 51st harmonic at $50490\,\text{Hz}$ will not be visible, since it is greater than $f_c/2$.

Due to the presence of harmonics beyond $f_c/2$, we are witnessing the phenomenon of aliasing, attributable to the presence of spurious components that "mirror" each other in the spectrum. The first of these aliasing components will therefore be present at

$$f_c/2 - (51 \cdot f_0 - f_c/2) = f_c - 51 \cdot f_0 = 49510\,\text{Hz}.$$

It can be noticed that the frequency at which the 50th harmonic is expected is actually the first mirrored spurious harmonic due to aliasing, as shown in Fig. 3.11. Obviously, the other spurious components will also be reflected on the screen from right to left.

To compare the results obtained using different window functions for the FFT, the span and the central frequency are set to display only the fundamental harmonic. The same square wave of step 1 (with frequency $f_0 = 990\,\text{Hz}$) is generated.

The ***Rectangular*** window, shown in Fig. 3.12, is useful for transient signals and signals in which there is an integral number of cycles in the time record. In fact, the rectangular window produces the maximum resolution in frequency, the main lobe is very narrow, and the lateral lobes decay more slowly, introducing a greater spectral dispersion.

The *Blackman–Harris* window, shown in Fig. 3.13, reduces time resolution compared to the rectangular window, but it improves the capability to detect smaller impulses due to lower secondary lobes (provides minimal spectral leakage).

The *Flat Top*, shown in Fig. 3.14, is the best for making accurate amplitude measurements of frequency peaks. In fact, the Flat Top window has the flat main lobe, which is wider than the previous ones. The side lobes are very small, and consequently, so is the spectral dispersion.

The *Hanning* window, shown in Fig. 3.15, is useful for obtaining good frequency resolution and for general-purpose use. This is the default window. In fact, Hanning's window has a double-width main lobe compared to the rectangular window and therefore a frequency resolution that worsens, but a more accentuated side lobe decay that reduces spectral dispersion.

For clarity, the resulting FFTs are shown in MATLAB graphs in Fig. 3.16. Finally, the four resulting FFTs are superimposed and compared in Fig. 3.17.

Table 3.1 summarizes the results of the measured amplitudes of the harmonics close to the fundamental.

3.2.2 Harmonic Distortion of a Signal Generated by an Oscillator

Equipment:

1. a digital oscilloscope with FFT functionality (Keysight InfiniiVision DSO X-2012A);
2. a function generator (HP 33120A);
3. DC power supply (HP E3631A) [3];
4. PCB containing an oscillator as shown in Fig. A.10 of Appendix A;
5. connecting cables and adapters as necessary.

Tasks:

1. using the software function of the given oscilloscope, visualize the spectrum of the signal generated by an oscillator;
2. measure the amplitude of the harmonics;
3. calculate the total harmonic distortion (THD) of the generated signal.

An oscillator is an input-free circuit that generates a periodic output signal. In the case of the proposed experience, it is a sinusoidal oscillator. All the oscillators are characterized by a positive reaction that prevents the circuit from reaching an equi-

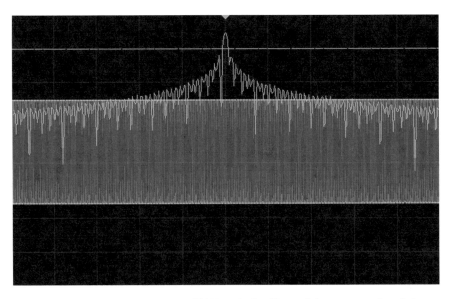

Fig. 3.12 FFT of a square wave at $f_0 = 990$ Hz, calculated by applying a rectangular window

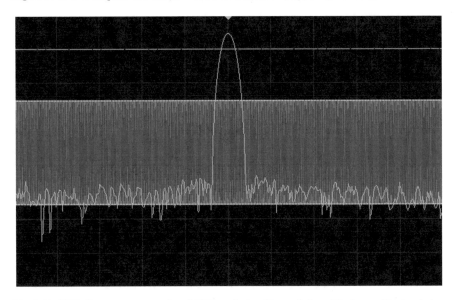

Fig. 3.13 FFT of a square wave at $f_0 = 990$ Hz, calculated by applying a Blackman–Harris window

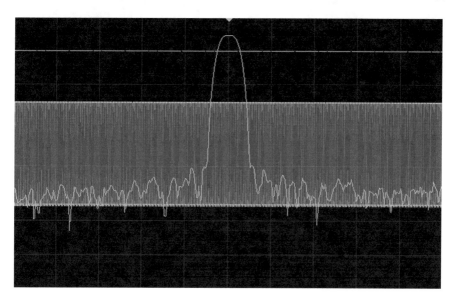

Fig. 3.14 FFT of a square wave at $f_0 = 990\,\text{Hz}$, calculated by applying a Flat Top window

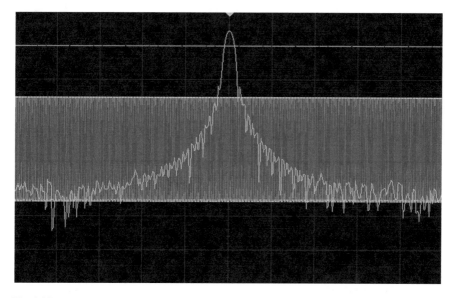

Fig. 3.15 FFT of a square wave at $f_0 = 990\,\text{Hz}$, calculated by applying a Hanning window

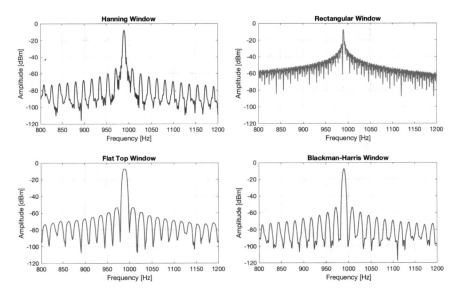

Fig. 3.16 MATLAB graphs of the FFT of a square wave at $f_0 = 990\,$Hz, calculated through four window functions

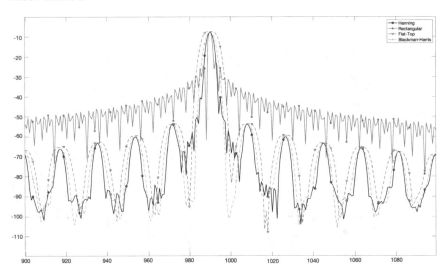

Fig. 3.17 Comparison of the FFT results of a square wave at $f_0 = 990\,$Hz, obtained using the different window functions

Table 3.1 Measured amplitudes of the harmonics for the four window functions

f (Hz)	Hanning dB$_V$	Flat Top dB$_V$	BlackMan dB$_V$	Rectangular dB$_V$
988.09	−13.21	−7.59	−10.42	−61.78
988.86	−9.61	−7.37	−8.46	−12.8
989.62	−7.9	−7.35	−7.48	−7.78
990.38	−7.89	−7.35	−7.47	−7.79
991.14	−9.56	−7.37	−8.43	−13.16
991.91	−13.12	−7.58	−10.38	−26.26

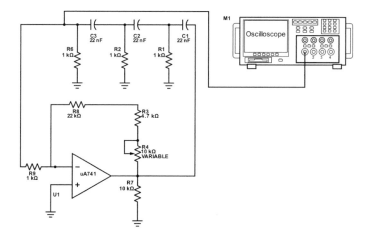

Fig. 3.18 Schematic of the oscillator

librium condition: the output signal oscillates, therefore, in a harmonic way (sine wave).

The oscillator analyzed in the following experience is called a *phase-shift oscillator*; see Fig. 3.18. It consists of a reaction network that introduces a phase shift of 180°. In particular, this phase shift is obtained from the series of three CR cells, each of which introduces a phase shift of 60°.

Since there is an operational amplifier, which is an active element, it is necessary to supply it with ±15 V through a DC power supply (the HP E3631A).

Considering that $C = 22$ nF, $R = 1$ kΩ, the passive network determines the phase shift of 180° at the theoretical frequency, which can be determined according to the following equation:

$$f = \frac{1}{2\pi \, RC\sqrt{6}} = \frac{1}{2\pi \cdot 22 \cdot 10^{-9} \cdot 10^3 \cdot \sqrt{6}} = 2.95 \,\text{kHz} . \tag{3.9}$$

The procedural steps are as follows:

1. Power the op-amp and slightly distort the signal and visualize its spectrum, as shown in Fig. 3.19.
2. Measure the oscillation frequency f_{os_m} and compare it with the calculated theoretical value:

$$f_{os_m} = \frac{1}{2\pi RC\sqrt{6}} \simeq 2.78\,\text{kHz}. \tag{3.10}$$

3. The *Span* and *Center* setting parameters of the oscilloscope have been set to display the first 10 harmonics, as shown in Fig. 3.20. To perform the FFT, a rectangular window has been set. In order to emphasize the effect of aliasing, the sampling frequency has been set at $f_c = 100\,\text{kHz}$, so that the folding is at 50 kHz. In this case, the number of harmonics (n_h) before the folding is

$$n_h = \frac{f_c/2}{f_0} = \frac{50\,\text{kHz}}{2.78\,\text{kHz}} = 17, \tag{3.11}$$

and therefore the first appreciable aliasing component is the 18th, indicated with an arrow in Fig. 3.21

$$f_c/2 - (18 \cdot f_0 - f_c/2) = f_c - 18 \cdot f_0 \simeq 49960\,\text{Hz}. \tag{3.12}$$

4. Measure the amplitude values (expressed in dBv) of the displayed higher-order harmonics up to the 10th ($2 \cdot f_{os_m}, 3 \cdot f_{os_m}, 4 \cdot f_{os_m} \ldots$), as shown in Table 3.2.
5. Calculate the ***total harmonic distortion (THD)*** of the oscillator signal. To perform this calculation, it is necessary to convert dBV to V. Finally, it is possible to calculate THD:

$$THD = 100 \cdot \frac{\sqrt{\sum_{i=2}^{10} E_i^2}}{E_1} \simeq 0.88\% \tag{3.13}$$

where E_1 is the voltage of the fundamental and E_i the voltage of the ith harmonic.

3.3 Basic Measurements Using the GSP-730 Gw Instek Analog Spectrum Analyzer

In this section, we propose different laboratory experiences using a low-cost swept-spectrum SA, namely the GSP-730 Gw Instek. This is a low-cost piece of equipment, mainly developed to fulfill the demands of RF communication education, as stated by the manufacturer [2]. The uncertainty specifications for frequency and amplitude measurements for the GSP-730 Gw Instek are reported in Fig. 3.22.

Therefore, from the reported specifications, it is assumed that the worst-case uncertainty for amplitude measurements is 2 dB, and for frequency measurements, it is 3.

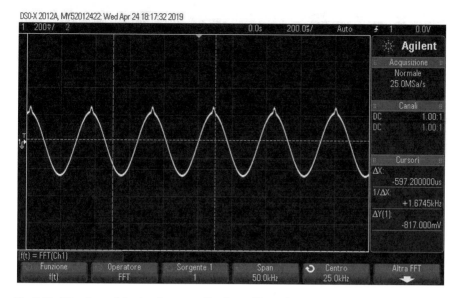

Fig. 3.19   Waveform of the signal generated by the oscillator after applying an intentional distortion

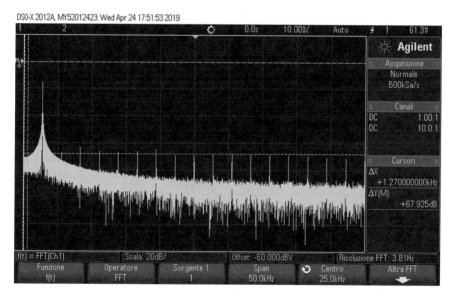

Fig. 3.20   FD visualization (using the FFT function of the oscilloscope) of the first ten harmonics of the oscillator signal

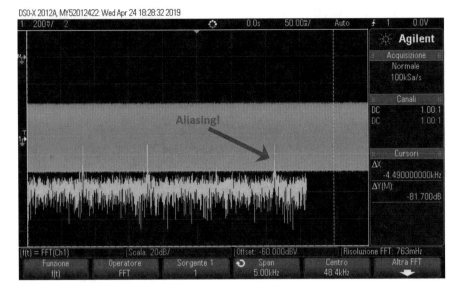

Fig. 3.21 Aliasing

Table 3.2 Measurement results of the amplitude values of the harmonics

Harmonics (kHz)	Amplitude (dBV)
2.78	−15.00
5.56	−66.88
8.34	−62.50
11.12	−65.63
13.90	−66.25
16.68	−65.00
19.46	−63.75
22.24	−68.75
25.02	−68.75
27.80	−68.13

3.3.1 Visualization of a Sinusoidal Signal and Its Spurious Components

Equipment:

1. a function generator (Hewlett-Packard 33120A);
2. an analog spectrum analyzer (GSP-730 GwInstek);
3. connecting cables and adapters as necessary.

GSP-730			
FREQUENCY	**Frequency Range**	Setting Range	150kHz ~ 3GHz
	Center Frequency	Setting Resolution	0.1MHz
		Accuracy	within ±50kHz (frequency span : 0.3GHz ~ 2.6GHz, 20 ±5°C)
	Frequency Span	Setting range	1MHz ~ 3GHz
		Accuracy	within ±3% (frequency span : 0.3GHz ~ 2.6GHz, 20 ±5°C)
	Resolution Bandwidth	Setting Range	30KHz, 100KHz, 300KHz,1MHz
	SSB Phase Noise		-85dBc/Hz (typical, 500kHz offset, RBW : 30kHz, Sweep time : 1.5s, Span : 1MHz@1GHz)
	Inherent Spurious Response		less than -45dBc-40dBm Ref. Level (typical less than -50dBc)
AMPLITUDE	**Reference Level**	Input Range	+20 ~ -40dBm
		Accuracy	Within ±2dB (1GHz) ; SPAN : 5MHz
		Unit	dBm, dBV, dBμV
	Average Noise Level		≤ -100dBm (typical, center frequency : 1GHz RBW : 30kHz)
	Frequency Characteristic		within ±3.0dB@300MHz ~ 2.6GHz
			within ±6.0dB@80 ~ 300MHz, 2.6 ~ 3GHz
	Input	Input Impedance	50Ω
		Input VSWR	less than 2.0@input att≥10dB
		Input damage level	+30dBm (CW average power), 25VDC
		Input connector	N connector
SWEEP	**Sweep Time**	Setting Range	300ms ~ 8.4s, auto (not adjustable)
		Accuracy	within ±2% (frequency span : full span)

Fig. 3.22 GSP-730 GwInstek uncertainty specifications for frequency and amplitude measurements

Tasks:

1. Use the SA to display a sinusoidal signal with frequency $f = 10\,\text{MHz}$ and peak-to-peak amplitude equal to 3 V.
2. Measure the frequency and amplitude of the fundamental.
3. Display the first 10 harmonics and set the "marker delta" command to measure the frequency and amplitude ranges between the fundamental and subsequent harmonics. Manually optimize (if necessary) the resolution of the IF filter.

The procedural steps are as follows:

1. Set the function generator to 10 MHz and 3 V_{pp}.
2. Set the parameters of the spectrum analyzer (frequency axis, amplitudes, and span) appropriately in order to display the signal spectrum and measure the frequency and amplitude of the fundamental harmonic, as shown in Fig. 3.23. The frequency and amplitude of the fundamental component can be measured with the peak search function

$$f = (10.00 \pm 0.30) \text{ MHz}, \qquad (3.14)$$

$$V_0 = (13.9 \pm 2.0) \text{ dBm}. \qquad (3.15)$$

3. With regard to the visualization of the first 10 harmonics, Fig. 3.24 shows the useful frequency range to be considered.

Because of the specifications of the given SA, even after optimizing the instrument settings, only one harmonic component after the fundamental can be measured. In fact, the noise floor is higher than the amplitude of the subsequent harmonics. The frequency of the first harmonic after the fundamental and the amplitude difference are

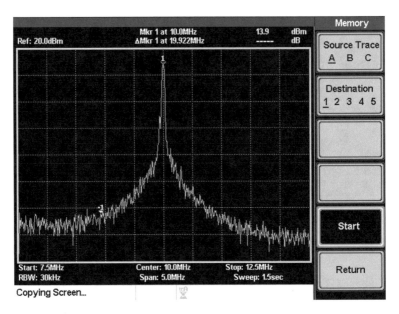

Fig. 3.23 Visualization of the fundamental at 10.0 MHz

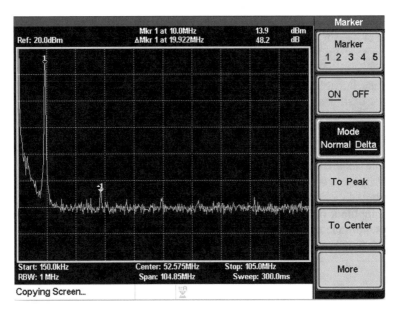

Fig. 3.24 Visualization of the frequency spectrum in the range 150.0 kHz–105.0 MHz. It can be seen that only one harmonic component is correctly displayed (the noise floor is higher than the amplitude of the subsequent harmonics)

$$f = (19.99 \pm 0.60) \text{ MHz},\tag{3.16}$$

$$\Delta V = V_0 - V_1 = 48.2 \text{ dB} \pm 4 \text{ dB}.\tag{3.17}$$

This uncertainty is due to the fact that the measurement is carried out as the difference between two measures.

3.3.2 Measurements of Two Sinusoidal Signals Having a Similar Amplitude and "Very Close in Frequency"

Equipment:

1. two function generators (Hewlett-Packard 33120A);
2. an analog spectrum analyzer (GSP-730 GwInstek);
3. a T-junction connector;
4. connecting cables and adapters as necessary.

Tasks:

1. Display two sinusoidal signals with the same peak-to-peak amplitude and with very close frequencies (f_1 and f_2).
2. Then, by varying the frequency of either signal, decrease the frequency separation ($\Delta f = |f_1 - f_2|$) between the two signals. The task is to identify the value of Δf that still allows you to identify the two separate fundamental harmonic components (optimize the RBW of the IF filter as necessary).

First of all, it should be noticed that the T-connector introduces an impedance mismatch: each generator has an impedance of 50 Ω, but with the T-connection, the generators are in parallel, and thus the total impedance is 25 Ω. Therefore, the voltage read on the analyzer differs from that set on the function generator.

The procedural steps are as follows:

1. Set one function generator to $f_1 = 10.0$ MHz and 3 V_{pp1} and the other one to $f_2 = 10.5$ MHz and 3 V_{pp2}. Supply the input to the spectrum analyzer via a T-connector and visualize them as shown in Fig. 3.25.
2. Now, to determine the minimum frequency interval Δf so that it is possible to "correctly" display the two fundamental harmonic components, we decrease the f_2 value. At the same time, it is necessary to optimize the RBW of the IF filter.

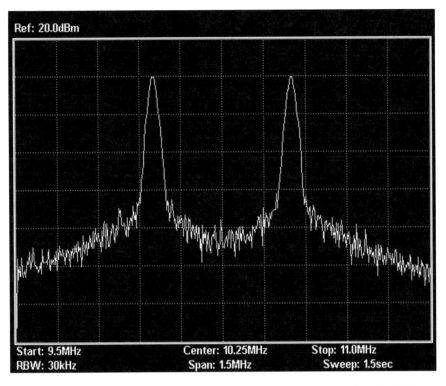

Fig. 3.25 Visualization of the fundamental components of the signals at $f_1 = 10.0$ MHz and $f_2 = 10.5$ MHz

It can be seen that the minimum difference Δf to display both signals correctly is 30 kHz (as shown in Fig. 3.26), which is the RBW value.

3.3.3 Measurements of Two Sinusoidal Signals Having Different Amplitudes

Equipment:

1. two function generators (Hewlett-Packard 33120A);
2. an analog spectrum analyzer (GSP-730 GwInstek);
3. a T-junction connector;
4. connecting cables and adapters as necessary.

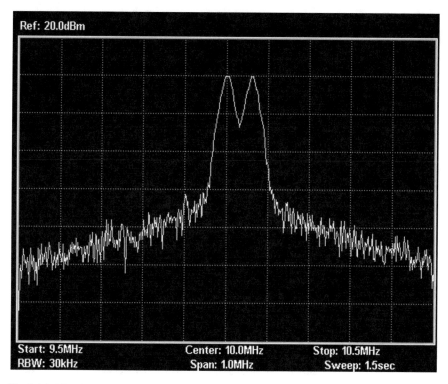

Fig. 3.26 Visualization of the fundamental components of the signal at $f_1 = 10.000$ MHz and of the signal at $f_2 = 10.030$ MHz, with Δf equal to 30 kHz

Tasks:

1. Generate and display two sinusoidal signals with the same frequency ($f_1 = f_2$), but with very different amplitudes (e.g., one signal with amplitude 0 dBm and the other signal with amplitude 20 dBm). On the SA, you will see that the lower-amplitude signal is "hidden" by the higher-amplitude signal.
2. Then, by varying the frequency of the lower-amplitude signal, identify the minimum value of Δf (where $\Delta f = |f_1 - f_2|$) that allows you to see both the sinusoidal signals on the SA display (optimize the RBW of the IF filter as necessary).

The procedural steps are as follows:

1. Set both function generators to the same frequency equal to 10 MHz but with different levels of amplitude, namely, the first one to 20 dBm and the second one to 0 dBm. Supply the input to the spectrum analyzer via a T-connector and visualize them as shown in Fig. 3.27.

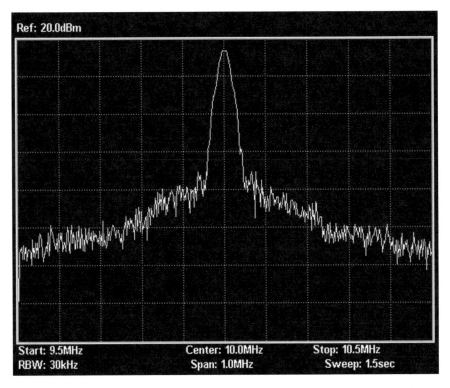

Fig. 3.27 Visualization of two sinusoidal signals at 10 MHz, one with amplitude 0 dBm and the other one with amplitude 20 dBm. The 20 dBm signal clearly "hides" the signal with lower amplitude

2. Properly set the frequency axis on the analyzer and the span. Moreover, set the reference for the amplitude to 20 dBm. Then with the generator, vary the frequency of the lower-amplitude signal until you obtain the correct display, always optimizing the RBW of the IF filter.

It is observed experimentally that it is possible to discriminate the signal at lower amplitude (0 dBm) only by moving away about 38 kHz with respect to the signal of 20 dBm centered at 10 MHz. This is the result of the fact that until this condition occurs, the signal with smaller amplitude is overwhelmed by the frequency response of the IF filter itself. This is shown in Fig. 3.28. Now the frequency difference between the two sinusoidal signals is such that the smaller signal "emerges" from the "bell-shaped" response of the IF filter. This limit, in addition to the IF filter bandwidth, also depends on the amplitude of the smaller signal. In fact, f_1 is the frequency of the largest signal, f_2 is the frequency at which the largest signal has an amplitude equal to the smallest signal; k is the difference between these two frequencies. Let a be the frequency separation between the two signals. To see the smaller signal, it is necessary that $k < a$, as shown in Fig. 3.29. Hence, the smaller the difference

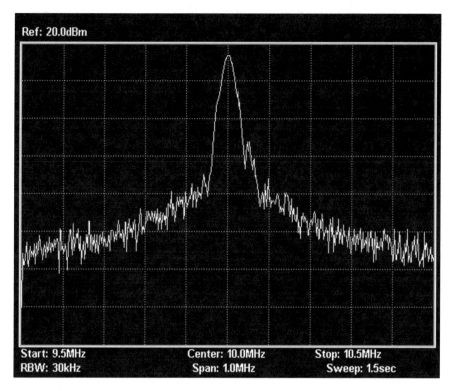

Fig. 3.28 Visualization of two sinusoidal signals, one with amplitude 0 dBm and the other with amplitude 20 dBm, after varying the frequency of the 0 dBm signal

Fig. 3.29 Comparison between the amplitudes and the separation in frequency of two tones

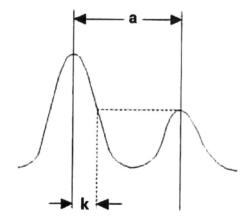

between the amplitudes, the lower the frequency at which we can distinguish the two components.

3.4   Measurements Using an HP E4411B Analog Spectrum Analyzer

In this section, similarly to the previous case, the simultaneous visualization of two sinusoidal signals is proposed, initially with the same amplitude (to analyze the frequency resolution capability), and subsequently with very different amplitudes (to observe selectivity capability).

The objective is to determine the minimum frequency separation between the two tones such that they can be suitably distinguished.

In the end, the signal-to-noise ratio and harmonic distortion are calculated. The uncertainties relating to frequency and amplitude measurements with this instrument are reported in Fig. 3.30: Therefore, from the reported specifications, it is assumed that the worst-case uncertainties for amplitude and frequency measurements, respectively, are

$$U_f = (f \cdot (255 \cdot 10^{-7})) + 0.75\% \cdot span + 15\% \cdot RBW + 10\,\text{Hz}\,, \qquad (3.18)$$

$$U_A = 1.1\,\text{dB}\,. \qquad (3.19)$$

It is worth noting that in contrast to the previous SA, the one considered in this section is characterized by better metrological performance.

Frequency reference error	±[(aging rate x time since last adjustment) + settability + temperature stability]
Frequency readout accuracy	(start, stop, center, marker) = ±((frequency indication x frequency reference error) + SP + 15% of RBW + 10 Hz)
Aging rate	±2 x 10⁻⁶/year
Tempurature stability	±5 x 10⁻⁶
Settability	±5 x 10⁻⁷
Span coefficient (SP)	0.75% x span

(a) *Frequency range*

Absolute 9 kHz to 3 GHz	±0.5 dB

(b) *Frequency response*

Overall amplitude accuracy	±(0.6 dB + absolute frequency response)

(c) *Absolute amplitude accuracy*

Fig. 3.30 Spectrum analyzer specifications

3.4.1 Visualization of a Sinusoidal Signal and Its Spurious Components

Equipment:

1. A function generator (Hewlett-Packard 33120A);
2. an analog spectrum analyzer (Hewlett-Packard E4411B);
3. connecting cables and adapters as necessary.

Tasks:

1. optimize the settings of the SA to display a sinusoidal signal with frequency equal to 10 MHz and peak-to-peak voltage $V_{pp} = 3$ V;
2. optimize the settings of the SA to display simultaneously the first 10 harmonics. Measure the frequency and amplitude ranges between the fundamental and subsequent harmonics. (Manually optimize, if necessary, the resolution of the IF and video filters.)

The procedural steps are as follows:

1. Set the function generator to $f_0 = 10$ MHz and 3 V_{pp}.
2. By appropriately setting the parameters of the SA, the signal spectrum is displayed as shown in Fig. 3.31. The amplitude of the fundamental is measured through the search peak function.
3. By displaying simultaneously the first 10 harmonics, as shown in Fig. 3.32, only the first five harmonics are visible, since the other harmonics are below the noise floor. Therefore, by appropriately adjusting the IF and video filters, it is advisable to display and measure separately each individual harmonic (modifying the span accordingly). By appropriately manipulating the IF and video filters and modifying the span, we can visualize individually up to the seventh harmonic above the fundamental. Even with the best setting of the SA, the additional higher-order harmonics are strongly attenuated and therefore comparable to the noise floor.

Tables 3.3 and 3.4 summarize the distances in frequency and the difference in amplitude of each harmonic of order higher than the fundamental f_0.

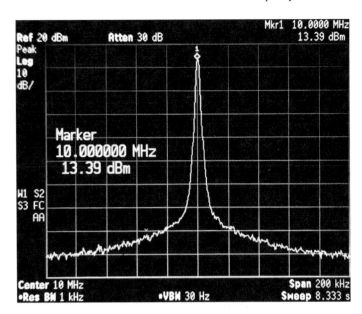

Fig. 3.31 Visualization of the fundamental component of the sinusoidal signal at $f = 10\,\text{MHz}$

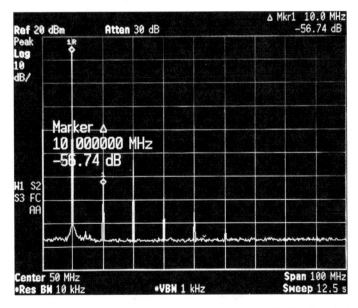

Fig. 3.32 Visualization of the first ten harmonic components of the 10 MHz sinusoidal signal

Table 3.3 Harmonics of high orders, distance in frequency. The SA settings are also reported

# of harmonic –	f (MHz)	Span (MHz)	RBW (kHz)	U_f (MHz)	$U_{r,f}$ (%)
1	10.000	100	10	0.038	0.38
2	20.00	100	10	0.76	3.8
3	30.00	100	10	0.76	2.5
4	40.00	100	10	0.76	1.9
5	50.00	100	10	0.76	1.5
6	60.00	100	10	0.76	1.3
7	70.000	4	1	0.030	0.044
8	80.00000	0.1	1	0.00077	0.00096

Table 3.4 Measured amplitude values of the first eight harmonics. Uncertainty values are also reported

Harmonics –	A (dBm)	U_A (dB)
0	13.41	1.1
1	−43.33	1.1
2	−32.83	1.1
3	−48.34	1.1
4	−55.01	1.1
5	−60.71	1.1
6	−71.57	1.1
7	−71.23	1.1

3.4.2 Measurements of Two Sinusoidal Signals Having a Similar Amplitude and "Very Close in Frequency"

Equipment:

1. two function generators (Hewlett-Packard 33120A);
2. analog spectrum analyzer (Hewlett-Packard E4411B);
3. T-junction connector for BNC;
4. connecting cables and adapters as necessary.

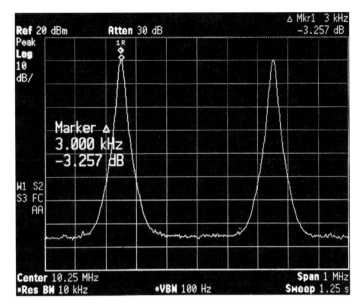

Fig. 3.33 Visualization of the spectra of the sinusoidal signals at $f_1 = 10.0$ MHz and $f_2 = 10.5$ MHz

Tasks:

1. Set the first function generator to 10.0 MHz and 3 V_{pp1} and the second one to 10.5 MHz and 3 V_{pp2};
2. Supply the input to the spectrum analyzer via a T-connector and display the two signals on the SA;
3. Determine the minimum frequency interval Δf (by appropriately varying the frequency of the signal with the second generator) so that it is possible to display the two fundamental harmonic components, optimizing the RBW of the IF filter and the video filter.

The procedural steps are as follows:

1. Set the first function generator to $f_1 = 10.0$ MHz and 3 V_{pp1} and the second one to $f_2 = 10.5$ MHz and 3 V_{pp2}.
2. Supply the input to the spectrum analyzer via a T-connector and visualize them, as shown in Fig. 3.33.

By reducing f_2, the signal spectrum will progressively appear closer to the spectrum of the signal with frequency f_1. As f_2 is reduced, the span and the RBW should be suitably adjusted in order to display the two signals in the optimal conditions.

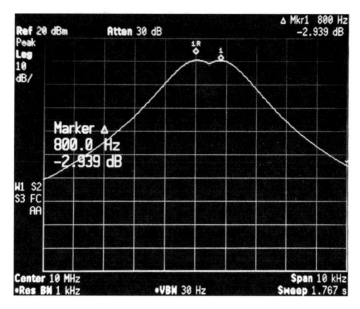

Fig. 3.34 Spectra of the sinusoidal signal at $f_1 = 10.0\,\text{MHz}$ and a sinusoidal signal separated by $\Delta f = 800\,\text{Hz}$

By proceeding as described above, it turns out that (for the SA used) the minimum difference in frequency to display both signals adequately is 800 Hz, as shown in Fig. 3.34. This is possible due to the fine adjustment of the RBW of the IF filter and also of the frequency bandwidth of the video filter (VBW).

It is important to note that with this configuration, the two generators and the spectrum analyzer have an impedance of 50 Ω, so the equivalent impedance seen by the latter is 25 Ω, and therefore the amplitude displayed on the analyzer will be 2/3 (linear scale) of that actually imposed on the function generators.

3.4.3 Measurements on Two Sinusoidal Signals Having Different Amplitudes

Equipment:

1. two function generators (Hewlett-Packard 33120A);
2. analog spectrum analyzer (Hewlett-Packard E4411B);
3. T-junction connector for BNC;
4. connecting cables and accessories as necessary.

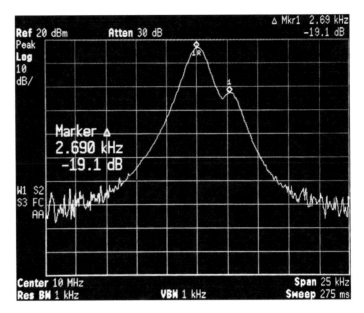

Fig. 3.35 Minimum distance in frequency to discriminate the two signals

Tasks:

Determine the minimum frequency interval Δf (by appropriately varying the frequency of the signal with the second generator) so that it is possible to display the two fundamental harmonic components, optimizing the RBW of the IF filter and the video filter of two signals with same frequency and very different amplitudes.

The procedural steps are as follows:

1. Set both function generators to the same frequency equal to 10 MHz but with different levels of amplitude, namely, the first one to 20 dBm and the second one to 0 dBm.
2. Supply the signal from the two generators to the input port of the spectrum analyzer via a T-connector.
3. Properly set the frequency axis on the analyzer and the span. Notice that, with these settings, you cannot discriminate the presence of the two different signals; in fact, the 0 dBm signal is covered by the 20 dBm signal at the same frequency. Moreover, set the reference for the amplitude to 20 dBm and the vertical scale to 10 dB/div. Then with the generator, vary the frequency of the smallest signal in amplitude until you can distinguish the presence of the two signals on the SA display, always optimizing the RBW of the IF filter and the VBW of the video filter.

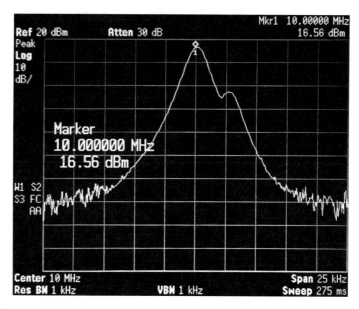

Fig. 3.36 Amplitude displayed on the spectrum analyzer

It is possible to discriminate the signal at a lower amplitude (0 dBm) by moving away from about 2.69 kHz with respect to the signal of 20 dBm centered at 10 MHz, as shown in Fig. 3.35. This is possible because with this analyzer it is possible to make a finer adjustment of the RBW, which allows us to go down to 1 kHz, (with the previous analyzer the limit was 30 kHz for the RBW) combined with the effect of the video filter also with a 1 kHz VBW.

One should also note the aforementioned effect regarding the display on the analyzer of the 2/3 (linear scale) amplitudes that are set on the function generators. Specifically, with respect to the 20 dBm signal at 10 MHz, approximately 16.56 dBm will be displayed on the analyzer screen, as shown in Fig. 3.36.

3.4.4   Signal-to-Noise Ratio Measurement

Equipment:

1. two function generators (Hewlett-Packard 33120A);
2. analog spectrum analyzer (Hewlett-Packard E4411B);
3. two coaxial cables (BNC-BNC type cable);
4. T-junction connector for BNC.

Tasks:

Measure the signal-to-noise ratio in two cases: **(a)** first by generating a sinusoidal signal through the internal generator of the SA and considering the internal noise of the SA. **(b)** Then the two function generators will be used, one to generate a sinusoidal signal and the other to generate noise.

(a) In this case, only the SA is used. The HP E4411B is in fact equipped with an internal generator. The procedural steps are as follows:

1. Turn on the internal 50 MHz alignment signal of the analyzer and use it as the signal being measured.
2. Set the center frequency to 50 MHz and the span to 1 MHz.
3. Set the reference level to −10 dBm.
4. Set the attenuation to 40 dB.
5. Place a marker on the peak of the signal.
6. Put the *delta marker* in the noise at a specified offset of 200 kHz (assuming that this point is out of the band of the signal of interest).
7. Use the noise marker to view the results of the signal-to-noise measurement out of the band of the signal of interest. This quantity is measured in dB/Hz, as shown in Fig. 3.37.

   At this point, we can easily compute the in-band noise-to-signal ratio N/S:

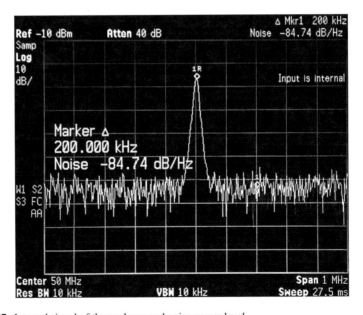

Fig. 3.37 Internal signal of the analyzer and noise power level

$$N/S = -84.74 \text{ dB/Hz} + 10 \cdot \log(200 \text{ kHz}) = \frac{-31.7297 \text{ dB}}{200 \text{ kHz}}. \tag{3.20}$$

(b) The setup used in these experiments includes:

- Two function generators *Hewlett-Packard 33120A*;
- analog spectrum analyzer *Hewlett-Packard E4411B*;
- two coaxial cables (BNC-BNC type cables);
- T-junction connector for BNC.

The procedural steps are as follows:

1. Set one function generator to 10 MHz and 3 V_{pp}.
2. Set the second function generator to generate white noise with an amplitude of 0 dBm.
3. Use the T-junction connector, and connect both the function generators to the input of the spectrum analyzer. Place a marker on the peak of the signal.
4. Put the *delta marker* in the noise at a specified offset of 200 kHz (assuming that this point is out of the band of the signal of interest).
5. Use the noise marker to view the results of the signal-to-noise measurement out of the band of the signal of interest. This quantity is measured in dB/Hz.
6. Repeat steps 3, 4, and 5 after increasing the noise power level to 10 dBm.

This is the N/S ratio from Fig. 3.38:

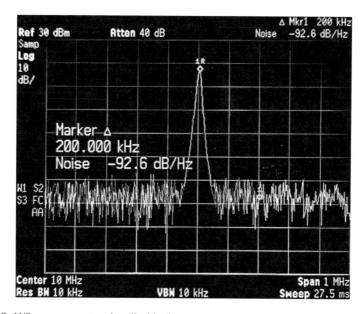

Fig. 3.38 N/S measurement as described in (b)

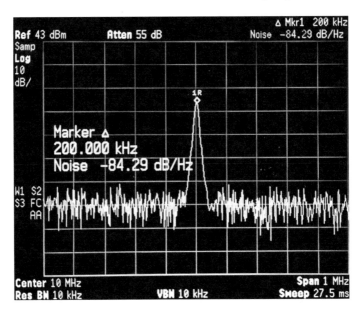

Fig. 3.39 N/S measurement as described in (b) after increasing the power noise level to 10 dBm

$$N/S = -92.6 \text{ dB/Hz} + 10 \cdot \log(200 \text{ kHz}) = \frac{-39.5897 \text{ dB}}{200 \text{ kHz}}. \qquad (3.21)$$

N/S after the increase in the noise power level from Fig. 3.39:

$$N/S = -84.29 \text{ dB/Hz} + 10 \cdot \log(200 \text{ kHz}) = \frac{-31.2797 \text{ dB}}{200 \text{ kHz}}. \qquad (3.22)$$

3.4.5 Measurement of Total Harmonic Distortion (THD)

Equipment:

1. function generator (Hewlett-Packard 33120A);
2. analog spectrum analyzer (Hewlett-Packard E4411B);
3. T-junction connector for BNC;
4. connecting cables and adapters as necessary.

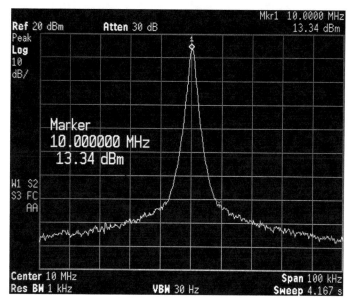

Fig. 3.40 3 V_{pp} sinusoidal signal from the function generator at 10 MHz

Tasks:

Calculate the total harmonic distortion (THD) of a sinusoidal signal (case **a**) and of a triangular signal (case **b**).

(a) Sinusoidal signal.

The procedural steps are as follows:

1. Set the function generator to a sinusoidal signal at 10 MHz and 3 V_{pp}. Connect the function generator to the spectrum analyzer, and display the signal as shown in Fig. 3.40.
2. Set the frequency range of the SA, in order to display the first ten harmonics. In the experience reported herein, only the first seven harmonics could be distinguished from the noise floor. Now, measure the amplitude of each of the harmonics, and compute the *THD*.

The results of the measurements are reported in Table 3.5.

After the conversion of the measured amplitudes of each harmonic in volts, the *THD* can be calculated as:

$$THD = 100 \cdot \frac{\sqrt{\sum_{i=2}^{7} E_i^2}}{E_1} \approx 0.45\%, \tag{3.23}$$

Table 3.5 Measurement of the amplitudes of the harmonics

Harmonics (MHz)	Amplitude (dBm)
10.00	13.34
20.00	−43.28
30.00	−34.12
40.00	−50.41
50.00	−55.56
60.00	−64.07
70.00	−70.36

Fig. 3.41 Uncertainty dB versus linear scale

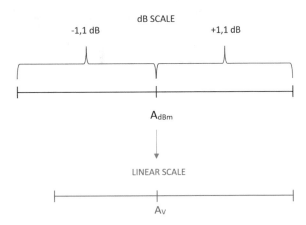

where E_1 is the voltage amplitude of the fundamental, and the E_i are the voltage amplitudes from the second harmonic to the seventh. It is now useful to introduce a specific note about the evaluation of the uncertainties affecting the amplitude, frequency, and harmonic distortion measurements, and in particular, about the transformation of uncertainty expressed in dB in a corresponding linear value. To do this, it is assumed that f is the frequency read on the analyzer, A_{dBm} the amplitude measured on a logarithmic scale, and A_V the amplitude converted to a linear scale. Considering that $A_V = \sqrt{10^{\frac{A_{dBm}}{10}} \cdot 50\,\Omega \cdot 1\,\text{mW}}$, then we have

$$U_{A_{dB}} = 0.6\,\text{dB} + 0.5\,\text{dB},\tag{3.24}$$

$$U_{A_V} = (A_V \cdot 10^{\frac{U_{A_{dB}}}{20}}) - A_V,\tag{3.25}$$

$$U_{r,A_V} = \frac{U_{A_V}}{A_V} \cdot 100.\tag{3.26}$$

Considering that the uncertainty from the instrument's specifications is given in dB, the corresponding uncertainty range is asymmetric, having the positive half-interval greater than the negative one, as illustrated in Fig. 3.41.

Fig. 3.42 Uncertainty range, worst case

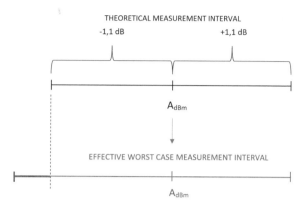

THEORETICAL MEASUREMENT INTERVAL

-1,1 dB   +1,1 dB

A_{dBm}

EFFECTIVE WORST CASE MEASUREMENT INTERVAL

A_{dBm}

Fig. 3.43 Uncertainty range, propagation formula

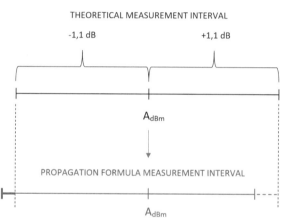

THEORETICAL MEASUREMENT INTERVAL

-1,1 dB   +1,1 dB

A_{dBm}

PROPAGATION FORMULA MEASUREMENT INTERVAL

A_{dBm}

As a matter of fact, for evaluating the absolute uncertainty corresponding to the linear scale, the upper bound can be considered (for a conservative evaluation), and consequently, the uncertainty can be found by subtracting the central value from the upper bound. This is a conservative method, since by converting the values back to a decibel scale, a slightly lower value is obtained for the negative bound of the uncertainty interval. This effect is highlighted in Fig. 3.42.

However, considering the expression in decibels as an indirect quantity, it is also admissible to calculate the uncertainty values through the well-known LPU, thus obtaining

$$U_A = \frac{\partial A_V}{\partial A_{dB}} \cdot U_{A_{dBm}} = \sqrt{1\text{mW} \cdot 50\Omega} \cdot 10^{\frac{A_{dBm}}{20}} \cdot \frac{ln(10)}{20} \cdot U_{A_{dB}} . \tag{3.27}$$

This approach, however, is less conservative than the previous one, since it is possible to note in Fig. 3.43 that a part of the uncertainty interval is uncovered.

Table 3.6 Frequency uncertainties

f (MHz)	Span (kHz)	RBW (kHz)	U_f (MHz)	$U_{r,f}$ (%)
10.00000	100	1	0.00099	0.0099
20.0000	100	1	0.0011	0.0053
30.0000	100	1	0.0012	0.0038
40.0000	100	1	0.0013	0.0031
50.0000	100	1	0.0013	0.0026
60.0000	100	1	0.0014	0.0023
70.0000	100	1	0.0015	0.0021

Table 3.7 Amplitude uncertainties

f (MHz)	A (dBm)	U_A (dB)	A_V (mV)	U_{A_V} (mV)	$U_{r,A}$ (%)
10.00000	13.3	1.1	1039	141	14
20.0000	−43.3	1.1	1.53	0.21	14
30.0000	−34.1	1.1	4.40	0.60	14
40.0000	−50.4	1.1	0.675	0.092	14
50.0000	−55.6	1.1	0.373	0.051	14
60.0000	−64.1	1.1	0.140	0.019	14
70.0000	−70.4	1.1	0.0678	0.0092	14

Tables 3.6 and 3.7 summarize the results obtained for the uncertainties of the previous measurements both in frequencies and amplitudes of each harmonic component, respectively.

A specific case concerning the calculation of the uncertainty relates to the THD measurements, for which the LPU has been applied:

$$U_{THD} = \sum_{i=1}^{7} \left| \frac{\partial THD}{\partial E_i} \right| \cdot U_{E_i} . \qquad (3.28)$$

Calculating the partial derivatives of the function with respect to each E_i, the following equation is obtained

$$U_{THD} = \frac{1}{E_f} \sum_{i=2}^{7} \left| \frac{E_i}{\sqrt{\sum_{k=2}^{7} E_i^2}} \right| \cdot U_{E_i} + \left| -\frac{\sqrt{\sum_{k=2}^{7} E_i^2}}{E_f^2} \right| \cdot U_{E_f} , \qquad (3.29)$$

and thus for the value of $THD = 0.45$. It is also worth mentioning that the SA used has a function that allows the automatic evaluation of the THD, as shown in Fig. 3.44. It can be seen that the automatic measurement of THD is in good agreement with

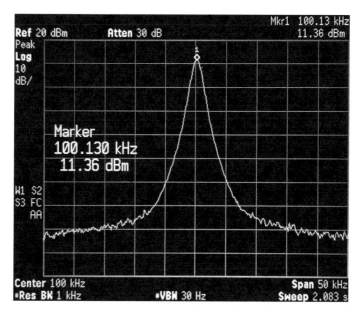

Fig. 3.44 Automatic *THD* measurement provided by the instrument

Fig. 3.45 3 V_{pp}-amplitude triangular signal at 100 kHz

the *THD* value calculated previously. It can also be noticed that the automatic *THD* evaluation allows one to measure the amplitudes of the harmonics up to the 10th.

(b) Triangular signal.

In this second case, a triangular signal with a frequency of 100 kHz and an amplitude of 3 V_{pp} is considered, as shown in Fig. 3.45. As predicted by Fourier's theorem, a periodic triangular wave is expected to have significant amplitude contributions corresponding to the odd harmonics, while the even ones are strongly attenuated.

Set the frequency range of the SA in order to display ten harmonics, and measure the amplitude of the harmonics that emerge from the noise floor. Results are reported in Table 3.8.

After the conversion of the measured amplitudes of each harmonic into volts, a *THD* approximately equal to 12.1% is obtained. Also in this case, the result can be compared with that obtained from the automatic measurement, as can be seen in Fig. 3.46, confirming a very good agreement.

Table 3.8 Measurement of the amplitudes of the harmonics for the triangular signal

Odd harmonics (kHz)	Amplitude (dBm)
100.00	11.36
300.00	−7.69
500.00	−16.52
700.00	−22.31
900.00	−26.63

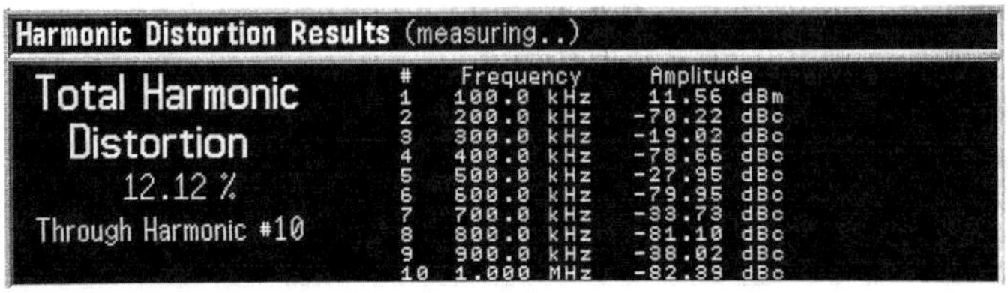

Fig. 3.46 Automatic *THD* measurement for a triangular signal

Table 3.9 Measured frequency values of the harmonics and corresponding uncertainty values

f (MHz)	Span (kHz)	RBW (kHz)	U_f (MHz)	$U_{r,f}$ (%)
100.00000	50	1	0.54	0.54
300.0000	50	1	0.54	0.18
500.0000	50	1	0.54	0.11
700.0000	50	1	0.55	0.078
900.0000	50	1	0.55	0.061

Table 3.10 Measured amplitude values of the harmonics and corresponding uncertainty values

f (MHz)	A (dBm)	U_A (dB)	A_V (mV)	U_{A_V} (mV)	$U_{r,A}$ (%)
100.00	11.36	1.1	827	112	14
300.00	−7.69	1.1	92	13	14
500.00	−16.52	1.1	33.4	4.6	14
700.00	−22.31	1.1	17.1	2.4	14
900.00	−26.63	1.1	10.4	1.5	14

Finally, the obtained results, including the evaluated uncertainty values, are sum-marized in Tables 3.9 and 3.10, and applying also in this case the LPU, a percentage uncertainty equal to 3.3% has been calculated for the *THD*.

3.5 Measurements of Modulated Signals

This section is dedicated to SA-based measurements on modulated signals, which are of fundamental importance especially in the telecommunications field. After a brief overview of the basic theory describing the main examples of modulations, related measurement experiments are reported.

3.5.1 Basic Theory on Modulations

Modulation varies the characteristics of the carrier signal proportionally to the instantaneous variations of the modulating signal:

- The *modulating signal* is the one that contains the information.
- The *carrier signal* has a higher frequency than the modulating one.

For example:

- In AM (amplitude modulation), the modulating signal varies, i.e., modulates, the carrier amplitude.
- In FM (frequency modulation), the modulating signal varies, i.e., modulates, the carrier frequency.

All the modulation techniques cause an increase in the bandwidth occupied by the carrier because they generate lateral bands.

Amplitude Modulation

Figure 3.47 shows the principle of amplitude modulation signal can be represented by the following expression:

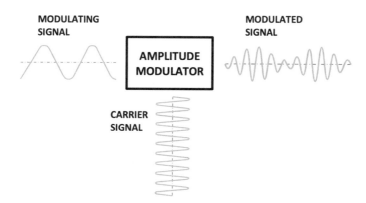

Fig. 3.47 Amplitude Modulation

Fig. 3.48 Ideal spectrum of the AM signal

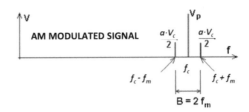

$$v(t) = A_c \times [1 + a \times m(t)] \times \cos(2\pi f_c t),\qquad(3.30)$$

where

- A_c determines the amplitude of the signal, and it is a constant;
- a is the modulation index, and it is between 0 and 1;
- $m(t)$ is the modulating signal;
- f_c is the carrier frequency.

If the modulating signal is sinusoidal, the expression becomes

$$v(t) = A_c \times \cos(2\pi f_c t) + A_c \times a \times m(t) \times \cos(2\pi f_c t)\qquad(3.31)$$
$$= A_c \times \cos(2\pi f_c t) + \tfrac{A_c a}{2} \times \cos[2\pi(f_c + f_m)] +$$
$$+ \tfrac{A_c a}{2} \times \cos[2\pi(f_c - f_m)].\qquad(3.32)$$

As shown in Fig. 3.48, the ideal spectrum of an AM signal has three spectral lines: the central line corresponds to the frequency of the carrier signal; the other two lines, on the left and on the right, are the side bands, caused by the modulation. The frequency distance of the side bands from the carrier is equal to the frequency of the modulating signal $f_m = f_c - f_{sb}$. The amplitude of the side bands A_s is proportional to the modulation index a:

$$A_s = \frac{a \times A_c}{2}.$$

Frequency Modulation
Frequency modulation (FM) consists in varying the frequency of the carrier, proportionally to the instantaneous value of the modulating signal, leaving the amplitude unchanged.

A frequency-modulated signal is represented by the following expression:

$$v(t) = A_c \times \cos\left(2\pi f_c t + k_f \int_i^t m(\tau) \times d\tau\right),\qquad(3.33)$$

where since the frequency is the derivative of the phase with respect to time, solving with respect to the phase yields

$$\theta(t) = k_f \int_i^t m(\tau) \times d\tau,\qquad(3.34)$$

Fig. 3.49 Ideal spectrum of the FM signal

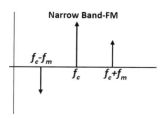

where $m(\tau)$ is the modulating signal and k_f is the deviation constant.

If the modulating signal is sinusoidal, the expression becomes

$$v(t) = A_c \times cos(2\pi f_c t + k_f \int_i^t A_m[cos 2\pi f_m t]d\tau) \tag{3.35}$$

$$= A_c \times cos(2\pi f_c t + \frac{k_f A_m}{2\pi f_m} sen[2\pi f_m t]) . \tag{3.36}$$

The modulation index β is

$$\beta = \frac{k_f A_m}{2\pi f_m} = \frac{\Delta f}{f_m} . \tag{3.37}$$

This index indicates how much the modulated signal varies with respect to its unmodulated form.

The quantity Δf is called the *frequency deviation* and represents the maximum difference between the instantaneous frequency of the modulated signal and the carrier frequency, while f_m is the frequency of the modulating signal.

Depending on the value assumed by the modulation index beta, two types of modulation can be distinguished:

1. **narrow band**;
2. **broadband**.

The narrow-band modulation condition occurs if $\beta \ll 1$. In fact, making the appropriate approximations, we obtain

$$v(t) = A_c \times cos(2\pi f_c t) + \frac{A_c \beta}{2} \times cos[2\pi(f_c + f_m)] + \frac{A_c \beta}{2} \times cos[2\pi(f_c - f_m)] , \tag{3.38}$$

and the ideal spectrum of the signal is represented in Fig. 3.49, noting that the difference between this and a modulated AM signal is the side-band phase.

When the modulation index is high, broadband angular modulation occurs. In the frequency domain, the signal occupies a larger bandwidth. In time domain, the expression of the modulated signal becomes:

$$v(t) = J_0(\beta) \cdot cos(2\pi f_c t) + \sum_{n=1}^{\infty} J_n(\beta)\{cos[2\pi(f_c - f_m)t] + cos[2\pi(f_c + f_m)t]\} \tag{3.39}$$

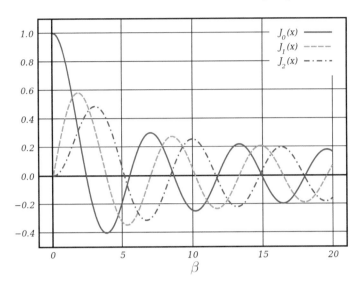

Fig. 3.50 Bessel function of the first kind

where $J_n(\beta)$ is a Bessel function of the first kind calculated in β.

Figure 3.50 shows that if β is very small, the predominant component is f_c, this means that the side bands have little contribution in the expression (3.39). Conversely, if β takes large values, then the higher-order Bessel functions take different values both for the carrier and for the side bands, some of which will certainly be present.

Ideally, the number of side bands is infinite, but their amplitude decreases as one moves away from the carrier frequency, and about 98% of all the power of an FM signal is within a bandwidth of

$$BW = 2(\Delta f + f_m). \tag{3.40}$$

Furthermore, increasing the modulating width increases the frequency deviation, but the total power transmitted does not change. However, the modulation band is widened as Δf increases.

Frequency Shift Keying (FSK)

Digital modulation FSK is a type of encoding in binary frequency-modulated form in which the modulating signal shifts the frequency of the output carrier from one to the other by two predetermined values. Moreover, being the width of the modulated signal independent of the modulating signal, this type of modulation is less sensitive to additive noise. The theoretical FSK modulated signal is shown in Fig. 3.51.

In this last section about spectral analysis, measurements are carried out on modulated signals.

Fig. 3.51 Sketch of
frequency shift keying (FSK)
modulation

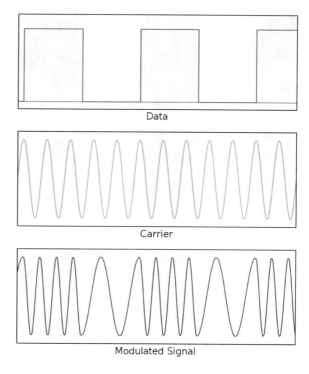

Data

Carrier

Modulated Signal

3.5.2 Measurements on AM Signals

Equipment:

1. an analog spectrum analyzer (Hewlett-Packard E4411B);
2. a digital oscilloscope (Keysight InfiniiVision DSO-X2012A);
3. a function generator (Hewlett-Packard 22120A): this function generator allows one to set different types of modulation with different modulation parameters;
4. connecting cables and adapters as necessary.

Tasks:

1. Use the function generator to generate an amplitude-modulated signal, with carrier frequency $f_c = 1$ MHz, modulating frequency $f_m = 20$ kHz, and modulation index $a = 100\%$. Display the frequency spectrum of the modulated signal.
2. Measure, through the SA, the values of f_c and the frequency values of the side bands (f_s). Also, calculate the values of f_m and a.
3. Using the oscilloscope, display in the time domain the amplitude-modulated signal of the previous step and comment on the waveform.
4. Repeat steps 1, 2, and 3, setting $a = 50\%$.

The procedural steps are as follows:

1. Select AM on the function generator, setting a *carrier frequency* (f_c) of 1 MHz, a *modulating frequency* (f_m) of 20 kHz, and a *modulation index* $a = 100\%$. The resulting amplitude-modulated signal is shown in Fig. 3.52.
2. Measure f_c, f_s, calculate f_m; measure A_c (carrier amplitude), A_s (amplitude of the modulating signal), calculate a, as reported in Fig. 3.52. The measurement results are summarized in Table 3.11 (Fig. 3.53):

$$f_m = f_s - f_c = 1.02013 \text{ MHz} - 1.00018 \text{ MHz} = 19.95 \text{ kHz}$$

$$A_s(dB) - A_c(dB) = 20\log\frac{a}{2} \implies a(\%) = 2 \cdot 10^{\frac{\Delta A_{dB}}{20}} = 99.59\% .$$

Using the oscilloscope, a modulated FM signal with modulation factor $a = 100\%$ will have the shape of Fig. 3.54 (for both signals the set amplitude is 3 V_{pp}, so we need to take into account the impedance mismatch between function generator and oscilloscope).

3. Repeat the same measurements but with modulation index $a = 50\%$. Figure 3.55 shows the signal in the frequency domain, with a marker at the side band.
   Here A_c, f_c, and f_s are the same as before, while A_s is different, and consequently $a = 2 \cdot 10^{-12.12/20} \approx 50\%$.
   Displaying the signal on the oscilloscope, as shown in Fig. 3.56, shows that there is a difference in excursion between maximum and minimum peaks with respect to the previous case. In this case, the modulation factor was halved and the $V_{max} - V_{min}$ excursion was also halved (ΔY in the figure). Regarding the uncertainties affecting the frequency and amplitude measurements, the same specifications and formulas of the previous experience are still valid. However, the uncertainty in the modulation index α must be considered:

$$U_{f_m} = U_{f_c} + U_{f_s} .$$

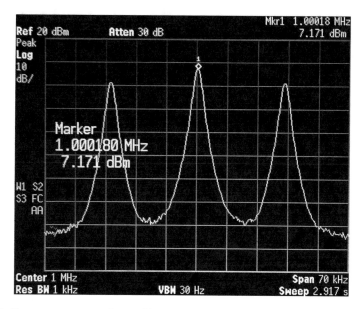

Fig. 3.52   Frequency spectrum of an amplitude-modulated signal with modulation index $a = 100\%$

Table 3.11   Measured parameters of the amplitude-modulated signal

f_c (MHz)	f_s (MHz)	A_c (dBm)	A_s (dBm)
1.00	1.02	7.17	1.11

In a linear scale, $a = 2 \cdot \dfrac{A_s}{A_c}$; so amplitudes in dB are converted to a linear scale: $A_{s_{LIN}} = 0.254\ V$ and $A_{c_{LIN}} = 0.511\ V$.

From the datasheet, we have $U_A = 1.1\ dB$; then $U_{A_s} = A_{s_{LIN}} - A_{s_{LIN}} \cdot 10^{\frac{1.1}{20}}$, $U_{A_c} = A_{c_{LIN}} - A_{c_{LIN}} \cdot 10^{\frac{1.1}{20}}$, and

$$U_a = U_{r,A} \cdot a = \left(\frac{U_{A_s}}{A_s} + \frac{U_{A_c}}{A_c} \right) \cdot a\ .$$

Tables 3.12, 3.13, and 3.14 summarize the uncertainty values for $a = 100\%$. Tables 3.15, 3.16, and 3.17 contain uncertainties for $\alpha = 50\%$.

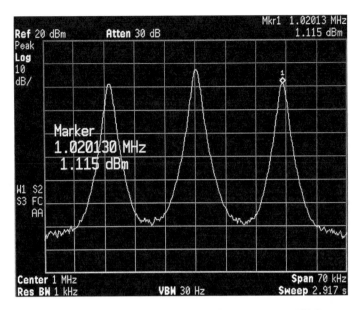

Fig. 3.53 Side band of modulated AM signal with modulation index $a = 100\%$

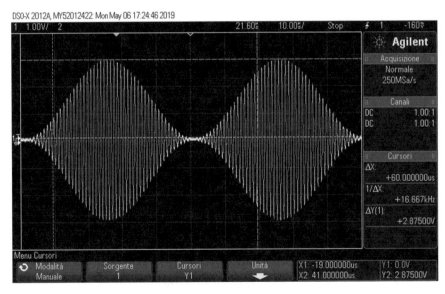

Fig. 3.54 Visualization in the time domain of an amplitude-modulated signal with modulation index $a = 100\%$

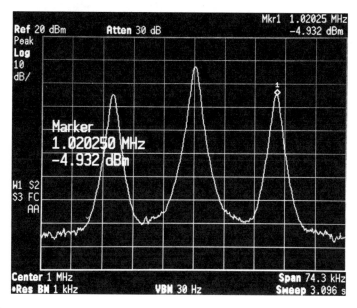

Fig. 3.55 Frequency spectrum of an amplitude-modulated signal with modulation index $a = 50\%$

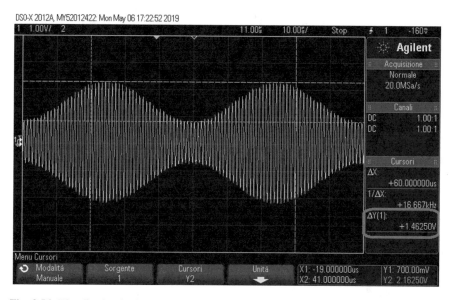

Fig. 3.56 Visualization in the time domain of an amplitude-modulated signal with modulation index $a = 50\%$

Table 3.12 Measured frequency values of the carrier and the side bands, with $a = 100\%$. The corresponding uncertainty values are also reported

Band –	f (MHz)	Span (kHz)	RBW (kHz)	U_f (MHz)	$U_{r,f}$ (%)
Carrier	1.00018	70	1	0.00070	0.070
Upper	1.02013	70	1	0.00070	0.068
Lower	0.98023	70	1	0.00070	0.071

Table 3.13 Measured amplitude values of the carrier and the side bands, with $a = 100\%$. The corresponding uncertainty values are also reported

Band –	f (MHz)	A (dBm)	U_A (dB)	A_V (V)	U_{A_V} (V)	$U_{r,A}$ (%)
Carrier	1.00018	7.2	1.1	0.511	0.069	14
Upper	1.02013	1.1	1.1	0.254	0.035	14
Lower	0.98023	1.1	1.1	0.254	0.035	14

Table 3.14 $a = 100\%$ uncertainties

Band –	ΔA (dB)	ΔA_{lin} –	$U_{r,\Delta A_{lin}}$ (%)	$U_{\Delta A_{lin}}$ –	a (%)	U_a (%)
Upper	−6.056	0.50	28	0.14	100	27
Lower	−6.069	0.50	28	0.14	100	27

Table 3.15 Frequency uncertainties with a = 50%

Band –	f (MHz)	Span (kHz)	RBW (kHz)	U_f (MHz)	$U_{r,f}$ (%)
Carrier	1.00018	70	1	0.00070	0.070
Upper	1.02013	70	1	0.00070	0.068

3.5.3   Measurements on FM Signals

Equipment:

1. an analog spectrum analyzer (Hewlett-Packard E4411B);
2. a digital oscilloscope (Keysight InfiniiVision DSO-X2012A);
3. a function generator (Hewlett-Packard 22120A), with an output impedance of 50 Ω (this function generator allows one to set different types of modulation with different modulation parameters);
4. connecting cables and adapters as necessary.

Table 3.16 Amplitude uncertainties with a = 50%

Band –	f (MHz)	A (dBm)	U_A (dB)	A_V (V)	U_{A_V} (V)	$U_{r,A}$ (%)
Carrier	1.00018	7.2	1.1	0.511	0.069	14
Upper	1.02013	−4.9	1.1	0.126	0.018	14

Table 3.17 a = 50% uncertainties

Band –	ΔA (dB)	ΔA_{lin} –	$U_{r,\Delta A_{lin}}$ (%)	$U_{\Delta A_{lin}}$ –	a (%)	U_a (%)
Upper	−12.12	0.25	28	0.067	50	14

Task:

1. Generate an FM signal (for example, with carrier frequency $f_c = 1\,\text{MHz}$, modulating frequency $f_m = 10\,\text{kHz}$) and display it on the SA and on the oscilloscope.
2. Use the SA to measure the carrier frequency (f_c), the side-band frequency (f_S), the amplitude of the carried (A_c), and the amplitude of the side bands (A_S).
3. Repeat the previous step for different values of the deviation frequency Δf in the 1–2 kHz frequency range.
4. Vary Δf and identify the value that leads to the narrow-band condition of the signal.
5. generate a, FM signal (for example, with carrier frequency $f_c = 100\,\text{kHz}$, modulating frequency $f_m = 4\,\text{kHz}$) to determine the condition of "cancellation" of the carrier.

The procedural steps are as follows:

1. Set the sinusoidal modulation FM with the function generator, setting a *carrier frequency* (f_c) of 1 MHz, a *modulating frequency* (f_m) of 10 kHz.
   The function generator allows one to set the frequency deviation $\Delta f = \beta \cdot f_m$, which represents the maximum difference between the instantaneous frequency of the modulated signal and the carrier frequency (i.e., the interval of variation of the carrier frequency).
2. Vary Δf in the range 1–2 kHz, so as to display the various spectral components relating to the side bands, and measure f_c, f_S, A_c, A_S, as shown in Fig. 3.57. Measures and uncertainties are reported in Tables 3.18, 3.19, 3.20, 3.21, 3.22, 3.23, and 3.24.
3. Determine the narrow-band condition (corresponding to the display of only two side bands), lowering the Δf value on the generator, as shown in Fig. 3.58. The value for which there is the narrow-band condition of the signal is $\Delta f = 300\,\text{Hz}$.

Table 3.18 FM measured data

Δf (kHz)	f_c (MHz)	f_m (kHz)	A_c (dBm)	A_{m1} (dBm)	A_{m2} (dBm)
1.00	1.00	10.00	13.26	−12.75	−43.98
1.50	1.00	10.00	13.23	−9.20	−37.44
2.00	1.00	10.00	13.19	−6.74	−32.53

Table 3.19 Frequency uncertainties with $\Delta f = 1$ kHz

Band –	f (MHz)	Span (kHz)	RBW (kHz)	U_f (MHz)	$U_{r,f}$ (%)
Carrier	1.00030	100	1	0.00092	0.092
Upper 1	1.01030	100	1	0.00092	0.091
Upper 2	1.02030	100	1	0.00092	0.090

If no phase information is available, as in this case (the spectrum analyzer used is not vectorial), the AM and FM signals appear indistinguishable in the frequency domain.

4. Set $f_c = 100$ kHz and $f_m = 4$ kHz to determine the condition of "suppression" of the carrier.

   When the modulation index β is high, in the frequency domain, the signal will occupy a greater bandwidth. The useful points for the cancellation of the carrier will be for $J_0(\beta) = 0$. For simplicity, only the first two points at which the Bessel function $J_0(\beta)$ has a zero are taken into account.

   The two corresponding β values are $\beta_1 \simeq 2.40$ and $\beta_2 \simeq 5.52$. So the corresponding frequency deviations are obtained as

$$\Delta f_1 = \beta_1 \times f_m = 2.40 \cdot 4 \text{ kHz} \approx 9.6 \text{ kHz},$$
$$\Delta f_2 = \beta_1 \times f_m = 5.52 \cdot 4 \text{ kHz} \approx 22.08 \text{ kHz}.$$

   The signal spectrum therefore becomes that shown in Fig. 3.59. Table 3.25 shows the measurements made in the carrier suppression conditions (for simplicity, the carrier and first side-band width were evaluated).

5. Display a modulated FM signal on the oscilloscope, setting the carrier f_c to 100 kHz, the modulating signal to 4 kHz, and the frequency deviation Δf to 25 kHz, as in Fig. 3.60. By using the horizontal cursors of the oscilloscope, it is possible to estimate the period of the sinusoidal signal at different time instants. It can be seen that, as a result of the FM, the frequency of the modulated signal varies, and so does its period.

Fig. 3.57 Frequency spectra
of an FM signal for different
values of Δf

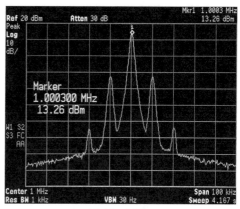

(a) $\Delta f = 1$ kHz.

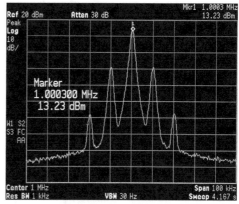

(b) $\Delta f = 1.5$ kHz.

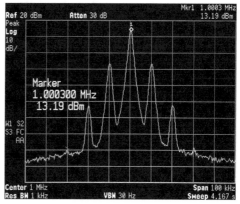

(c) $\Delta f = 2$ kHz.

Table 3.20 Amplitude uncertainties with $\Delta f = 1\,\text{kHz}$

Band	f	A	U_A	A_V	U_{A_V}	$U_{r,A}$
–	(MHz)	(dBm)	(dB)	(V)	(V)	(%)
Carrier	1.00030	13.3	1.1	1.03	0.14	14
Upper 1	1.01030	−12.8	1.1	0.0515	0.0070	14
Upper 2	1.02030	−44.0	1.1	0.00141	0.00020	14

Table 3.21 Frequency uncertainties with $\Delta f = 1.5\,\text{kHz}$

Band	f	Span	RBW	U_f	$U_{r,f}$
–	(MHz)	(kHz)	(kHz)	(MHz)	(%)
Carrier	1.00030	100	1	0.00092	0.092
Upper 1	1.01030	100	1	0.00092	0.091
Upper 2	1.02030	100	1	0.00092	0.090

Table 3.22 Amplitude uncertainties with $\Delta f = 1.5\,\text{kHz}$

Band	f	A	U_A	A_V	U_{A_V}	$U_{r,A}$
–	(MHz)	(dBm)	(dB)	(V)	(V)	(%)
Carrier	1.00030	13.2	1.1	1.03	0.14	14
Upper 1	1.01030	−9.2	1.1	0.0775	0.011	14
Upper 2	1.02030	−37.4	1.1	0.00300	0.00041	14

Table 3.23 Frequency uncertainties with $\Delta f = 2\,\text{kHz}$

Band	f	Span	RBW	U_f	$U_{r,f}$
–	(MHz)	(kHz)	(kHz)	(MHz)	(%)
Carrier	1.00030	100	1	0.00092	0.092
Upper 1	1.01030	100	1	0.00092	0.091
Upper 2	1.02030	100	1	0.00092	0.090

Table 3.24 Amplitude uncertainties with $\Delta f = 2\,\text{kHz}$

Band	f	A	U_A	A_V	U_{A_V}	$U_{r,A}$
–	(MHz)	(dBm)	(dB)	(V)	(V)	(%)
Carrier	1.00030	13.2	1.1	1.02	0.14	14
Upper 1	1.01030	−6.7	1.1	0.1029	0.014	14
Upper 2	1.02030	−32.5	1.1	0.00528	0.00072	14

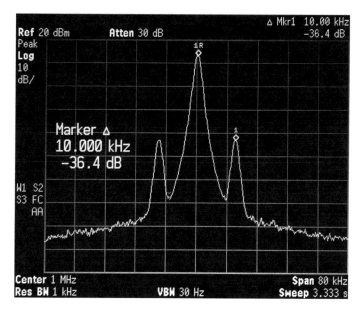

Fig. 3.58 Frequency spectrum of a narrow-band FM signal

Table 3.25 Measured data

Mod. Index	f_c	f_m	A_c	A_m
–	(kHz)	(kHz)	(dBm)	(dBm)
β_1	100.00	4.00	−33.92	7.8
β_2	100.00	4.00	−31.06	4.11

3.5.4 Measurements on an FSK Modulated Signal

Equipment:

1. a digital oscilloscope (Keysight InfiniiVision DSO-X2012A);
2. a function generator (Hewlett-Packard 22120A);
3. connecting cables and adapters as necessary.

Tasks:

1. Set an FSK modulation on the function generator and display the modulated signal on the oscilloscope.

Fig. 3.59 Visualization of
the spectrum of an FM signal
with suppression of the
carrier

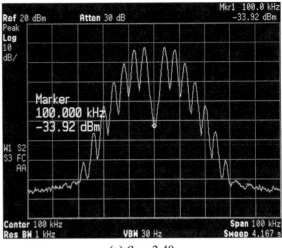

(a) $\beta_1 \simeq 2.40$.

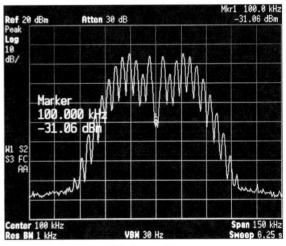

(b) $\beta_2 \simeq 5.52$.

The procedural steps are as follows:

1. Set the sinusoidal modulation FSK with the function generator, setting a *carrier frequency* (f_c) of 3 kHz and an amplitude of 5 V_{pp}.
2. Set also a *hop frequency* of 500 Hz and an *FSK rate* of 100 Hz.
3. Connect the function generator to the oscilloscope and display the FSK modulated signal, as shown in Fig. 3.61.

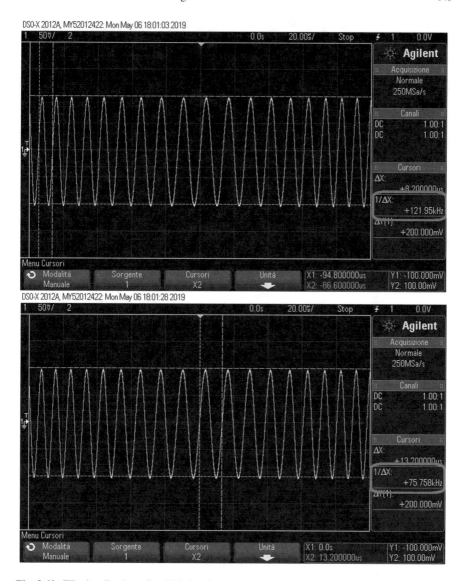

Fig. 3.60 TD visualization of an FM signal

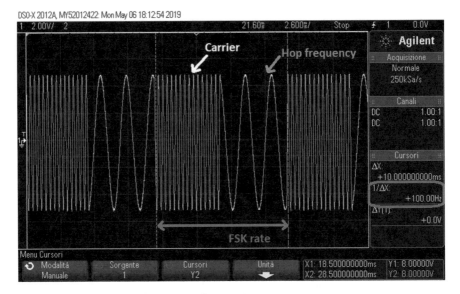

Fig. 3.61 Frequency-shift keying (FSK) with FSK rate of 100 Hz

References

1. Agilent ESA-L Series Spectrum Analyzers datasheet. Agilent Technologies, USA (December 5, 2012) https://literature.cdn.keysight.com/litweb/pdf/5989-9556EN.pdf?id=1504890
2. GSP-730 3GHz Spectrum Analyzer datesheet. GW Instek http://www.farnell.com/datasheets/2352591.pdf
3. Programmable DC Power Supplies datasheet. Keysight E363xA Series, USA (May 8, 2018) http://literature.cdn.keysight.com/litweb/pdf/5968-9726EN.pdf

Chapter 4
Reflectometric Measurements

Abstract Microwave reflectometry is a powerful measurement technique that can be effectively employed for a number of practical applications (e.g., for monitoring purposes, impedance variation evaluation, cable faults detection). This chapter addresses reflectometric measurements carried out in the time domain (TD). After a brief introduction to the theoretical aspects, several reflectometry-based laboratory experiments are presented concerning the identification and evaluation of unknown electric loads, the estimation of the wave propagation along cables, and the dielectric characterization of materials. As discussed later, some experiments are proposed twice, using two different types of equipment: (1) an oscilloscope and a function generator and (2) a dedicated instrument (i.e., a reflectometer).

4.1 Theoretical References

4.1.1 Time-Domain Reflectometry

Time-Domain Reflectometry (TDR) is a high-frequency measurement technique widely used for monitoring and diagnostics purposes [1–3]. When compared to other measurement techniques, TDR provides a more intuitive and direct view of the characteristics of the device under test [4].

Reflectometry relies on (1) propagating a suitable electromagnetic signal along a probe inserted in (or in contact with) the system under test and (2) analyzing the reflected signal.

In TD reflectometry, the test signal is generally a steplike signal or a pulse signal.

The incident and reflected voltage signal are acquired through an oscilloscope (although, often, a TDR measuring instrument includes both the generator and the acquisition section). Through an appropriate processing of the reflected signal, it is possible to retrieve the desired information on the monitored system.

© Springer Nature Switzerland AG 2020
A. Cataldo et al., *Basic Theory and Laboratory Experiments in Measurement and Instrumentation*, Lecture Notes in Electrical Engineering 663,
https://doi.org/10.1007/978-3-030-46740-1_4

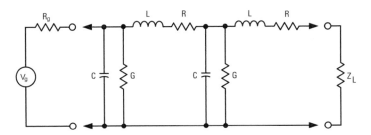

Fig. 4.1 The classical model for a transmission line

The classical transmission line is assumed to consist of a continuous structure of resistors, inductors, and capacitors, as shown in Fig. 4.1 [5, 7]. By studying this equivalent circuit, several characteristics of the transmission line can be determined.

If the line is infinitely long and R, L, G, and C are defined per unit length, then

$$Z_{in} = Z_0 = \sqrt{\frac{R + j\omega L}{G + j\omega C}}, \qquad (4.1)$$

where Z_0 is the characteristic impedance of the line. A voltage introduced at the generator will require a finite time to travel down the line to a point x. The phase of the voltage moving down the line will lag behind the voltage introduced at the generator by an amount β per unit length. Furthermore, the voltage will be attenuated by an amount α per unit length by the series resistance and shunt conductance of the line. The phase shift and attenuation are defined by the propagation constant γ, where

$$\gamma = \alpha + j\beta = \sqrt{(R + j\omega L)(G + j\omega C)}. \qquad (4.2)$$

The velocity at which the voltage travels down the line can be defined in terms of β:

$$v_\rho = \frac{\omega}{\beta} \text{ [unit length for second] }. \qquad (4.3)$$

In the general case, with v_c denoting the speed of light in vacuum and ϵ_r the dielectric constant of the medium, the relative propagation velocity of the wave in the medium is

$$v_\rho = \frac{v_c}{\sqrt{\epsilon_r}}. \qquad (4.4)$$

The propagation constant γ can be used to define the voltage and the current at any distance x down an infinitely long line by the relations

$$V_x = V_{in}e^{-\gamma x}, \quad I_x = I_{in}e^{-\gamma x}. \qquad (4.5)$$

The voltage and the current are related at every point by the characteristic impedance of the line

$$Z_0 = \frac{V_{in}e^{-\gamma x}}{I_{in}e^{-\gamma x}} = \frac{V_{in}}{I_{in}} = Z_{in} \,, \tag{4.6}$$

where V_{in} and I_{in} are the incident voltage and incident current respectively.

When the transmission line is finite and is terminated with a load (Z_L) whose impedance matches Z_0, the voltage and current relationships are satisfied by the preceding equations. If the load is different from Z_0, these equations are not satisfied unless a second wave is considered to originate at the load and to propagate back up the line toward the source. This reflected wave is energy that is not delivered to the load. Therefore, the quality of the transmission system is indicated by the ratio of this reflected wave (V_r) to the incident wave (V_i) originating at the source. This ratio is called the *voltage reflection coefficient*, denoted by ρ. It is related to the transmission line impedance by the equation

$$\rho = \frac{V_r}{V_i} = \frac{Z_L - Z_0}{Z_L + Z_0}. \tag{4.7}$$

The magnitude of the steady-state sinusoidal voltage along a line terminated in a load other than Z_0 varies periodically as a function of distance between a maximum and minimum value. This variation, called a *standing wave*, is caused by the phase relationship between incident and reflected waves. The ratio of the maximum and minimum values of this voltage is called the *voltage standing wave ratio (SWR)*, and is related to the reflection coefficient by the equation

$$\text{VSWR} = \frac{1 + |\rho|}{1 - |\rho|}. \tag{4.8}$$

For instance, assuming a discontinuity at a distance dx from the starting end, a reflected signal will reach the receiver after a time proportional to dx; see Fig. 4.2.

The reflected wave will be in phase opposition if in the discontinuity, the impedance is $Z_{dx} = 0$, while it will be in phase if $Z_{dx} = \infty$ with the incident wave.

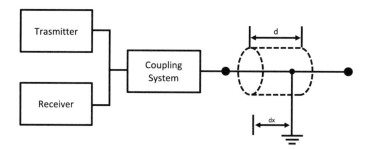

Fig. 4.2 Schematization of a reflectometric measurement technique

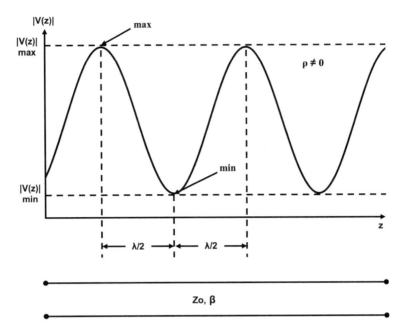

Fig. 4.3 Standing wave

This system is based on the definition of the aforementioned standing wave, where it is possible to associate a transmission line model to a cable and then evaluate the relative maxima and minima of the *standing wave diagram*, as shown in Fig. 4.3.

Standing waves are so called because the maxima and minima are anchored at fixed distances, although the wave propagates along the z-axis of the transmission line. It is quite simple to obtain $|V(z)_{max}|$ and $|V(z)_{min}|$ as follows:

$$|V(z)_{max}| = |V_i| \cdot |1 + |\rho|| \tag{4.9}$$

$$|V(z)_{min}| = |V_i| \cdot |1 - |\rho||. \tag{4.10}$$

From these quantities we evaluate the standing wave ratio (VSWR), which provides an indication of the impedance mismatch between the transmission line and its load. In the event that the load is adapted to the characteristic impedance, we have $SWR = 1$, since $\rho = 0$:

$$SWR = \frac{|V(z)_{max}|}{|V(z)_{min}|}. \tag{4.11}$$

Figure 4.4 shows a simplified sketch of a TDR-based measurement system, in particular, a simple case considering the TDR response of an electronic device characterized by an impedance Z_L.

As previously stated, the step generator produces an electromagnetic signal (often a steplike voltage signal) that is applied to the device under test (DUT), or more

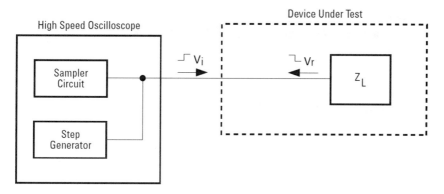

Fig. 4.4 TDR analysis scheme

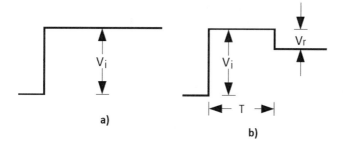

Fig. 4.5 **a** No reflection $V_r = 0$. **b** Reflection $V_r \neq 0$

generally, to the system under test. The EM signal travels along the transmission line at the velocity of propagation of the line. If the load impedance is equal to the characteristic impedance of the line, no wave is reflected; see Fig. 4.5(a). If a mismatch exists at the load, part of the incident wave is reflected. The reflected voltage wave will appear on the oscilloscope display algebraically added to the incident wave; see Fig. 4.5(b).

The reflected wave is readily identified, since it is separated in time from the incident wave. This separation can be determined by noting the length of the transmission system from the monitoring point to the mismatch. Denoting by D this length, we have

$$D = \frac{v_\rho \cdot T}{2}. \tag{4.12}$$

In particular, T is the round trip time, namely the transit time from the monitoring point to the mismatch and back again, as measured on the oscilloscope, while v_ρ is velocity of propagation in the cable.

Knowledge of V_i and V_r, as measured on the oscilloscope, allows us to determine the electric load in terms of the characteristic impedance of the transmission line, or vice versa.

4.1.2   Frequency-Domain Reflectometry

With the increased complexity and higher-frequency operation of devices, circuits, and systems, there is a critical need to accurately and efficiently characterize these advanced devices.

Understanding the magnitude and phase responses of a device under test (DUT) over a range of frequencies is a necessary step to do the following:

- fully characterize linear networks;
- effectively design impedance-matching networks;
- measure complex impedance;
- provide accurate models for computer-aided-engineering (CAE) circuit-simulation programs;
- apply error correction to enhance the accuracy of measurements.

Usually, these measurements are carried out through vector network analyzers (VNA) [6, 8]. Using power waves at various frequencies, the VNA measures the incident power (power sent into the DUT), reflected power, and transmitted power from each external connection of the device (port). VNAs can be used to measure the scattering parameters (S-parameters) of the DUT. S-parameters fully describe the input and output characteristics of the DUT based on power measurements. Other measurement parameters exist, such as H, Y, and Z parameters. But these parameters add more complications to the effective measurement of the electrical response of a DUT. These parameters can be mathematically derived from the S-parameters when needed. Physical quantities that can be evaluated starting from S-parameters include: return loss (power lost in reflection); voltage standing wave ratio (VSWR), which is the normalized power of the RF response over the frequency range); the transmission coefficient (the ratio of transmitted voltage to the incident voltage); and gain/attenuation (the ratio of signal power either gained or lost during transmission through the DUT). The description of the general architecture and of the functionalies of VNAs can be found in [6, 8].

4.2   TDR Measurements Using an Oscilloscope and a Waveform Generator

4.2.1   Experimental Characterization of Open and Short Circuits

Equipment:
1. a digital oscilloscope (Keysight InfiniiVision DSO X-2012A);
2. a function generator (Hewlett-Packard 33120A);
3. a T-junction with BNC connectors;
4. connecting cables and adapters as necessary.

Task:
Measure the reflectometric responses of known electric loads (open circuit, short circuit, matched load, $Z_0/2$) and comment on them with respect to the response expected from the theory.

4.2.1.1   Short Circuit Load (with and without Cable)

For this experiment, the short circuit is obtained in two ways: with and without a 50 cm-long portion of coaxial cable connected to one end of the Tee connector.

After connecting the T-junction to the input of the oscilloscope, one end is connected to the output of the function generator. The other end of the T-junction connector is short-circuited. The reflected wave subtracts itself from the incident one (in fact, $Z_L \simeq 0$) and the only visible effects are some fluctuations, as can be seen in Fig. 4.8.

From the theory, in the presence of a short circuit, the expected result is shown in Fig. 4.7.

The coaxial cable can be associated with a transmission line consisting of cells that also contain inductive components. These make the peaks more evident when the end of the coaxial cable is short-circuited, as in Fig. 4.6. In this case, the short circuit at the end of the coaxial cable was obtained by short-circuiting with an additional short metallic wire.

It is also important to observe how intertwining the short wire used for short-circuiting the inner and outer conductors at the end of the coaxial cable reduces the inductive effects (Fig. 4.9) (such as reducing the length of an inductive coil).

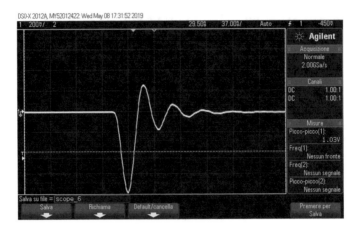

Fig. 4.6   Output waveform in the presence of a short circuit at the end of a coaxial cable

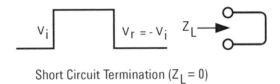

Fig. 4.7   Schematization of the expected TDR response in the presence of a short circuit

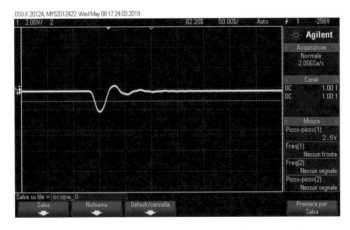

Fig. 4.8   Output waveform in the presence of a short circuit at the output of the T-connector

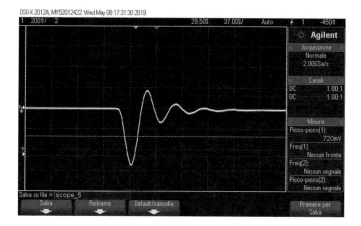

Fig. 4.9 Reduced inductive effects (in terms of reduction of the peak amplitude) as a result of better intertwining the short wire used for short-circuit at the end of the coaxial cable

Fig. 4.10 Schematization of the expected TDR response in the presence of an open circuit

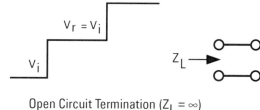

Open Circuit Termination ($Z_L = \infty$)

4.2.1.2 Open Circuit Load (with coaxial cable)

Figure 4.10 shows a schematization of the ideal TDR reflectogram corresponding to the case in which the signal propagates along a coaxial cable, with an open-ended termination.

The amplitude of the signal that is visualized on the oscilloscope is twice the one that provided by the generator. This happens because the oscilloscope is measuring the sum of the incident wave and the reflected wave (Fig. 4.11).

4.2.1.3 Matched Load

The schematization of the expected TDR response in presence of a matched load is shown in Fig. 4.12.

Connect a 50 cm coaxial cable to the T-junction and the 50 Ω load at its end. If the load impedance, Z_L, is equal to the characteristic impedance of the line, Z_0, the wave provided is displayed without alterations except for the peaks shown in Fig. 4.13; this shows that under these conditions, the cable does not affect the waveform, except for the addition of a capacitance effect caused by the cable, which explains the peaks.

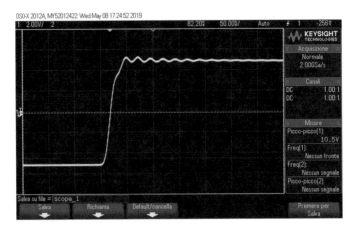

Fig. 4.11 Open circuit waveform acquired for the coaxial cable, with its end left open-circuited

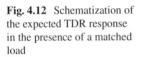
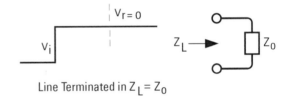

Fig. 4.12 Schematization of the expected TDR response in the presence of a matched load

Line Terminated in $Z_L = Z_0$

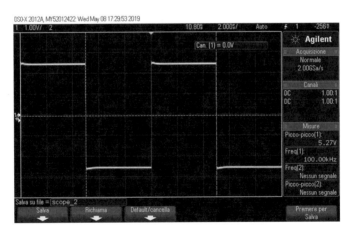

Fig. 4.13 Oscilloscope output with a 50 Ω load at the end of the cable

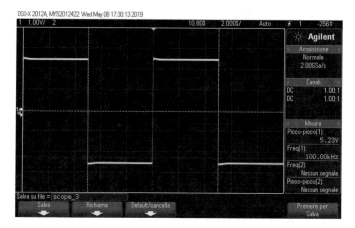

Fig. 4.14 Load without cable

Removing the coaxial cable, it is possible to apply directly to the T-junction connected to the oscilloscope a electric load with impedance $Z_L = 50\ \Omega$. The displayed wave is still the one generated, in fact, there are no reflections, and on the oscilloscope is visualized only the incident wave as it is generated, as in Fig. 4.14.

Before proceeding further, it is necessary to consider the following. The rise time of the signal generated by the function generator has a very long temporal duration compared to 200–300 ps, which is the typical value of a dedicated measuring instrument for TDR analysis. So since the oscilloscope has a sampling frequency that is low compared to the high-frequency analysis that has been carried out, it is impossible to discriminate the length of a 50 cm coaxial cable on the oscilloscope. Indeed, it was impossible to discriminate the typical staircase behavior that was expected, but only a single step was observed. In contrast, for tests with much longer cables, this kind of behavior would be apparent.

4.2.1.4   Load Z_L Equal to $Z_0/2$

Reflectometry allows us to assess not only the distance of the fault, but also its nature, e.g., an open circuit, an short circuit, etc. For example, it is also possible to recognize other fault conditions such as the failure impedance (or load Z_L) equal to half of the characteristic (Z_0). In this condition, the reflected wave is $1/3$ of the incident wave and is subtracted from it; see Fig. 4.15.

Then a T-junction is used with two 50 Ω terminations at its ends (in fact, the parallel of the two is equal to 25 Ω, that must be directly connected to the T-junction that connects the oscilloscope and the function generator. This situation generates the trend shown in Fig. 4.16.

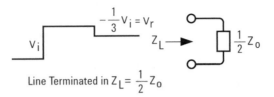

$$\text{Line Terminated in } Z_L = \frac{1}{2} Z_0$$

Fig. 4.15 Schematization of the expected TDR response in the presence of a load equal to 1/3 of the characteristic impedance

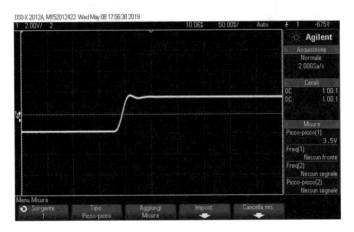

Fig. 4.16 Output waveform when the cable is terminated with a load equal to $Z_L = Z_0/2$

4.2.2 Measurement of Wave Propagation Speed on a Cable of Known Length

Equipment:
1. a digital oscilloscope (Keysight InfiniiVision Agilent DSO X-2012A);
2. a function generator (Hewlett-Packard 33120A);
3. coaxial cable of known length;
4. a T-junction connector.

Task:
Evaluate the wave propagation speed on the given coaxial cable.

This task consists in measuring the difference Δt between the wave V_i (incident wave) and V_r (reflected wave) of a cable of known length. This quantity represents

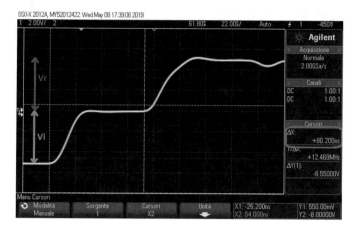

Fig. 4.17 Incident and reflected wave in a 7.8 m coaxial cable

the time taken by the voltage wave V_i to travel along the cable and return to the oscilloscope. For this experience, a cable of length 7.8 m has been used.

Using the horizontal cursors of the oscilloscope, the measured value of Δt is 80.20 ns; see Fig. 4.17.

Therefore, the propagation speed is given by

$$v_\rho = \frac{2 \cdot \text{cable length}}{\Delta t} = \frac{2 \cdot 7.8 \text{ m}}{80.20 \text{ ns}} \simeq 194514 \text{km/s}. \tag{4.13}$$

The length of the cable is multiplied by 2, since it is necessary to consider the round-trip route of the wave. To evaluate the dielectric constant of the insulating material (teflon) present between the coaxial cable conductors, it is possible to rewrite the propagation speed v_ρ as a propagation speed v_r relative to the speed of light v_c, that is, $v_r = \frac{v_\rho}{v_c} \simeq \frac{1}{\sqrt{\epsilon_r}} \simeq 2.37$.

4.2.3 Evaluation of the Unknown Length of a Coaxial Cable

Equipment:
1. a digital oscilloscope (Keysight InfiniiVision Agilent DSO X-2012A);
2. a function generator (Hewlett-Packard 33120A);
3. same type of coaxial cable as in the previous experience, but this time with unknown length;
4. a T-junction connector.

Fig. 4.18 Output waveform when a reel of coaxial cable is terminated with an open circuit

Task:
Evaluate the unknown length of the coaxial cable through reflectometric measurements.

For this experience, use the same type of coaxial cable as in the previous experience (or at least a coaxial cable with known dielectric characteristics). In the following, a reel of coaxial cable is considered and connected directly to the instruments without unrolling it. Consider a reel of cable of unknown length. Connect the reel of cable to the T-junction and leave its termination first in the open circuit condition and then shorted. The reel of cable used has identical dielectric characteristics to those of the one used in the previous experience and therefore the same propagation speed in the medium. Figure 4.18 is related to the reel with its distal end open-circuited.

Using the horizontal cursors of the oscilloscope, the measured Δt is 836.00 ns, and denoting by S_o the length of the reel of cable, we have

$$S_o = \frac{v \cdot \Delta t}{2} = \frac{194514 \text{ km/s} \cdot 836.00 \text{ ns}}{2} \simeq 81.31 \text{ m}. \tag{4.14}$$

To verify this measurement, repeat the same process for the short circuit termination, as in Fig. 4.19.

Also here it is quite simple to observe that Δt is more or less the same as in the open circuit case, specifically 838.00 ns, and so S_s (short-circuited length of the reel of cable) is

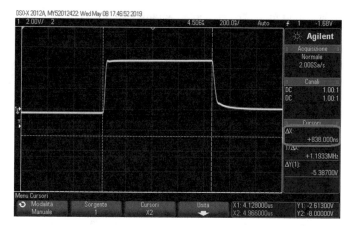

Fig. 4.19 Output waveform when a reel of coaxial cable is terminated with a short circuit

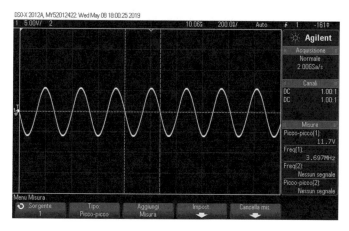

Fig. 4.20 Sinusoidal signal at 3.7 MHz

$$S_s = \frac{v \cdot \Delta t}{2} = \frac{194514 \text{ Km/s} \cdot 838.00 \text{ ns}}{2} \simeq 81.50 \text{ m} . \tag{4.15}$$

4.2.4 Frequency Modulation Echo System

In this practical experience, the concept of the stationary wave ratio (SWR) is addressed.

1. Set a sine wave on the function generator, frequency $f = 3.7$ MHz, peak-to-peak voltage $V_{pp} = 10$ V, and output impedance 50 Ω.

2. Connect the 7.8 m long coaxial cable, leaving it in open circuit.
3. Set the oscilloscope to display on one channel and multiple periods. The oscilloscope's display is shown in Fig. 4.20.
4. Change the frequency of the sinuosoidal signal to find two consecutive relative
   maxima, then annotate their frequencies in order to calculate the cable length
   (obviously, among the maxima there will always be relative minima). In this
   case, the first two consecutive maxima are at 11.40 MHz and around 100.00 kHz.
   So that the distance between the maxima is equal to $\Delta f = (11.40 - 0.10) =$
   11.30 MHz, the length x of the cable is

$$x = \frac{v}{2 \cdot \Delta f} = \frac{194514 \text{ km/s}}{2 \cdot 11.30 \text{ MHz}} \simeq 8.61 \text{ m}. \tag{4.16}$$

Clearly, the significant deviation from the reference value (7.8 m) is due to the
intrinsic limation of this measurement method, which provide qualitative results.

By repeating exactly the same previous steps on the cable reel used in the
previous experiment, the following frequencies relating to the first four consecutive relative maxima encountered were noted: $f_{m_1} = 1.20$ MHz, $f_{m_2} = 2.37$ MHz,
$f_{m_3} = 3.53$ MHz, and $f_{m_4} = 4.78$ MHz.

At this point, we can calculate the length:

$$x_1 = \frac{v}{2 \cdot (f_{m_2} - f_{m_1})} = \frac{194514 \text{ km/s}}{2 \cdot 1.17 \text{ MHz}} \simeq 83.13 \text{ m}, \tag{4.17}$$

$$x_2 = \frac{v}{2 \cdot (f_{m_3} - f_{m_2})} = \frac{194514 \text{ km/s}}{2 \cdot 1.16 \text{ MHz}} \simeq 83.84 \text{ m}, \tag{4.18}$$

$$x_3 = \frac{v}{2 \cdot (f_{m_4} - f_{m_3})} = \frac{194514 \text{ km/s}}{2 \cdot 1.25 \text{ MHz}} \simeq 77.80 \text{ m}. \tag{4.19}$$

By averaging these three distances, we obtain

$$x_{av} = \frac{x_1 + x_2 + x_3}{3} = 81.60 \text{ m}. \tag{4.20}$$

4.2.5 Identification of Unknown Electrical Loads

Equipment:
1. a digital oscilloscope (Keysight InfiniiVision Agilent DSO X-2012A);
2. a function generator (Hewlett-Packard 33120A);
3. coaxial cable of known length;
4. a T-junction junction connector;
5. a number of unknown electrical loads.

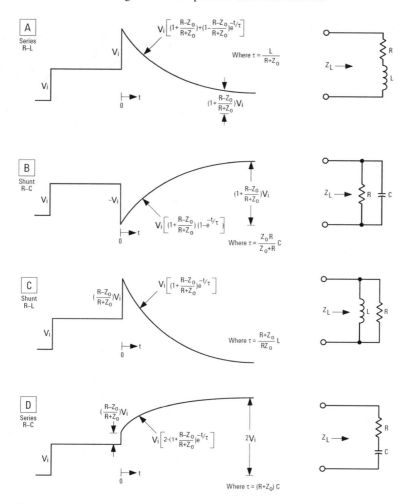

Fig. 4.21 Examples of possible types of unknown loads

Task:
Identify the behavior of eight unknown electrical loads through TDR measurements carried out through an oscilloscope.

In this last test the aim is to recognize eight unknown loads based on the behavior observed on the oscilloscope. The loads are attributable to the four cases presented in Fig. 4.21.

Figures 4.22, 4.23, 4.24, 4.25, 4.26, 4.27, 4.28, and 4.29 show the oscilloscope waveforms acquired for the following loads:

Fig. 4.22 Unknown load
number 1

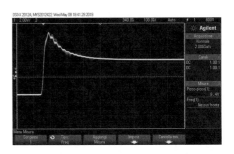

Fig. 4.23 Reflectogram of
the unknown electric load
number 2

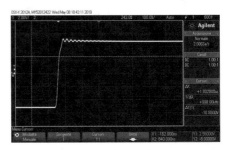

Fig. 4.24 Reflectogram of
the unknown electric load
number 3

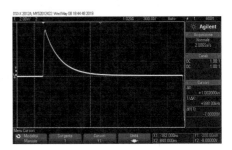

1. The first load in Fig. 4.22 can be associated to case **A**, $R - L$ series.
2. The second load in Fig. 4.23 cannot be associated to any of the previous ones, but can be seen as an open circuit.
3. The third load in Fig. 4.24 can be associated to case **C**, shunt $R - L$.
4. The fourth load in Fig. 4.25 can be associated to case **C**, shunt $R - L$.
5. The fifth load in Fig. 4.26 can be associated to case **B**, shunt $R - C$.
6. The sixth load in Fig. 4.27 can be associated to case **B**, shunt $R - C$.
7. The seventh load in Fig. 4.28 can be associated to case **D**, $R - C$ series.
8. The eighth load in Fig. 4.29 can be associated to case **D**, $R - C$ series.

These loads were used also in the subsequent experience with a dedicated TDR measuring instrument (i.e., a reflectometer that includes both the signal generator and the oscilloscope function), namely TDR100, where the values of the components have also been derived.

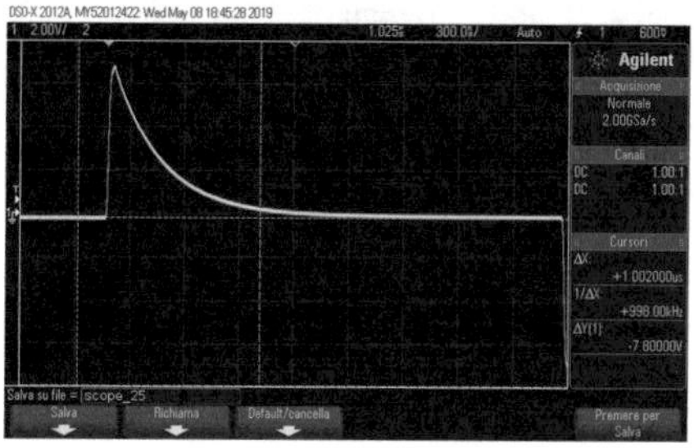

Fig. 4.25 Reflectogram of
the unknown electric load
number 4

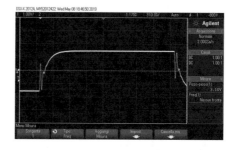

Fig. 4.26 Reflectogram of
the unknown electric load
number 5

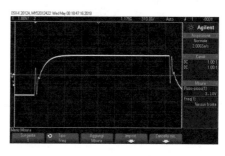

Fig. 4.27 Reflectogram of
the unknown electric load
number 6

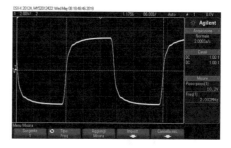

Fig. 4.28 Reflectogram of
the unknown electric load
number 7

Fig. 4.29 Reflectogram of
the unknown electric load
number 8

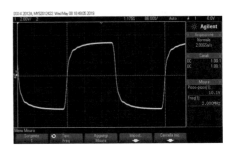

4.3 TDR Measurements Using a Reflectometer

4.3.1 Identification and Characterization of Unknown Electrical Loads

Equipment:
1. a reflectometer (TDR100) [9];
2. a T-junction connector;
3. a number of unknown electrical loads.

Task:
Identify the behavior of eight unknown electrical loads through TDR measurements and calculate their values.

In this experience, the same unknown loads used in the test with the oscilloscope are considered. Moreover, they have been associated to the same cases of Fig. 4.21 of Sect. 4.2.5. In order to compute the unknown values of the reactive components (capacitors and inductors), consider 5τ to be the time transient needed to obtain a constant value of the reflection coefficient. From this time value, the capacitance or inductance values were evaluated according to the case under consideration, as shown by the formulas reported in Fig. 4.21.

The steps of the procedure are listed only for the first case; as they are almost identical for the other cases.

1. Let D_b denote the distance value for which there is an approximately constant value of ρ, while D_a is the initial point where the effect of the load occurs.
2. Define ΔD to be the difference between D_b and D_a.
3. Assuming that the propagation takes place in vacuum, it is possible to obtain the time necessary to travel the ΔD distance as $\Delta t = \Delta D / v_c$ (v_c is the speed

Table 4.1 Evaluation of the first unknown load (Case A of Fig. 4.30)

Load type	ρ	Z_0 (Ω)	R (Ω)	D_a (m)	D_b (m)	ΔD (m)	Δt (5τ) (ns)	L (μH)
R-L series	0.05	50.00	55.26	1.39	114.39	112.99	376.65	7.93

of light). And since it has previously been assumed that after 5τ the transient is finished, impose $\Delta t = 5\tau$ (in nanoseconds).
4. At this point, all the quantities are known for obtaining R and L or C (in both cases the steps are very similar).

Explicit calculations:

$$\Delta D = D_b - D_a , \tag{4.21}$$

$$\Delta t = \frac{\Delta D}{v_c} = 5\tau , \tag{4.22}$$

$$R = Z_0 \cdot \frac{1+\rho}{1-\rho} , \tag{4.23}$$

$$\tau = \frac{\Delta t}{5} , \tag{4.24}$$

$$L = \tau \cdot (R + Z_0) \text{ or } L = \tau \cdot \frac{R \cdot Z_0}{R + Z_0} , \tag{4.25}$$

$$C = \tau \cdot \frac{R + Z_0}{R \cdot Z_0} \text{ or } C = \frac{\tau}{R + Z_0} . \tag{4.26}$$

Proceed with identification of the loads.

1. The reflectogram of the first load, shown in Fig. 4.30, can be associated with case **A** of Fig. 4.21, $R - L$ series. The related quantities are summarized in Table 4.1.
2. This second load, Fig. 4.31, cannot be associated to any of the previous ones, but can be associated with an open circuit.
3. This third load, Fig. 4.32, can be associated with case **C**, shunt $R - L$. The related quantities are summarized in Table 4.2.
4. This fourth load, Fig. 4.33, can be associated with case **C**, shunt $R - L$. The related quantities are summarized in Table 4.3.

5. This fifth load, Fig. 4.34, can be associated with case **B**, shunt $R - C$. The related quantities are summarized in Table 4.4.
6. This sixth load, Fig. 4.35, can be associated with case **B**, shunt $R - C$. The related quantities are summarized in Table 4.5.
7. This seventh load, Fig. 4.36, can be associated with case **D**, $R - C$ series. The related quantities are summarized in Table 4.6.
8. This eighth load, Fig. 4.37, can be associated with case **D**, $R - C$ series. The related quantities are summarized in Table 4.7.

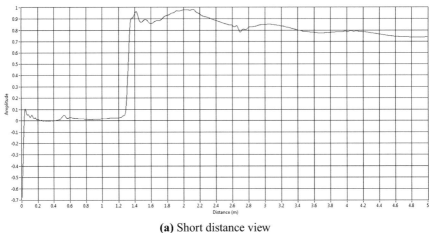

(a) Short distance view

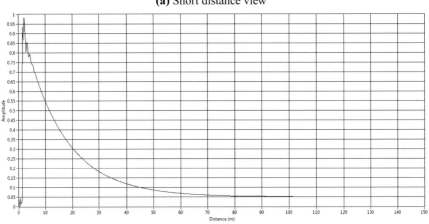

(b) Long distance view

Fig. 4.30 Unknown load number 1

Table 4.2 Case C load computation

Load type	ρ	Z_0 (Ω)	R (Ω)	D_a (m)	D_b (m)	ΔD (m)	Δt (5τ) (ns)	L (μH)
Shunt R-L	0.58	50.00	187.98	1.34	299.85	298.51	995.03	7.86

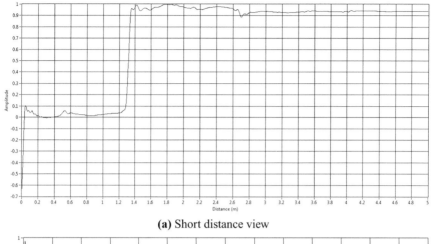

(a) Short distance view

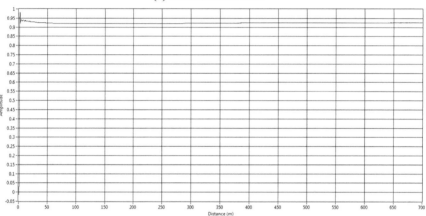

(b) Long distance view

Fig. 4.31 Unknown load number 2

Table 4.3 Case C load computation

Load type	ρ	Z_0 (Ω)	R (Ω)	D_a (m)	D_b (m)	ΔD (m)	Δt (5τ) (ns)	L (μH)
Shunt R-L	0.57	50.00	182.80	1.34	299.85	298.51	995.0	7.81

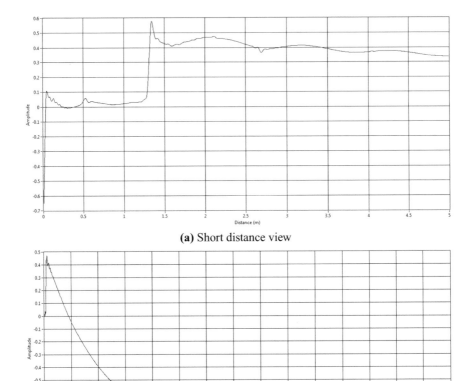

(a) Short distance view

(b) Long distance view

Fig. 4.32 Unknown load number 3

Table 4.4 Case B load computation

Load type	ρ	Z_0 (Ω)	R (Ω)	D_a (m)	D_b (m)	ΔD (m)	Δt (5τ) (ns)	C (nF)
Shunt R-C	-0.39	50.00	22.06	1.40	200.05	198.65	662.16	8.65

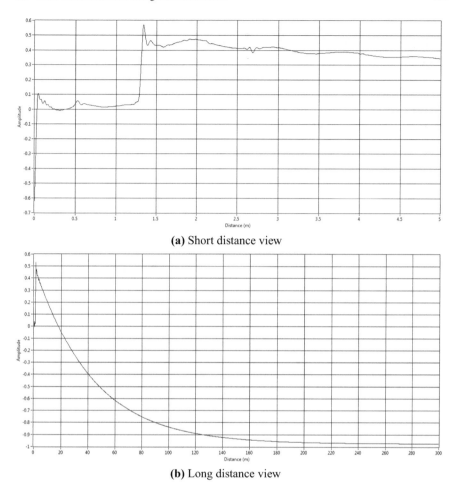

(a) Short distance view

(b) Long distance view

Fig. 4.33 Unknown load number 4

Table 4.5 Case B load computation

Load type	ρ	Z_0 (Ω)	R (Ω)	D_a (m)	D_b (m)	ΔD (m)	Δt (5τ) (ns)	C (nF)
Shunt R-C	−0.39	50.00	22.04	1.40	220.13	218.72	729.08	9.53

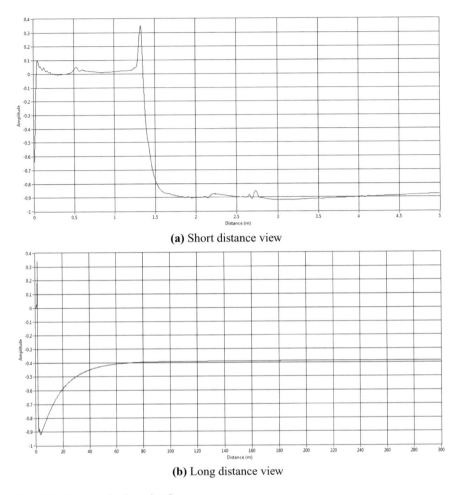

(a) Short distance view

(b) Long distance view

Fig. 4.34 Unknown load number 5

Table 4.6 Case D load computation

Load type	ρ	Z_0 (Ω)	R (Ω)	D_a (m)	D_b (m)	ΔD (m)	Δt (5τ) (ns)	C (nF)
R-C series	0.62	50.00	213.27	1.34	40.01	38.67	128.88	97.91

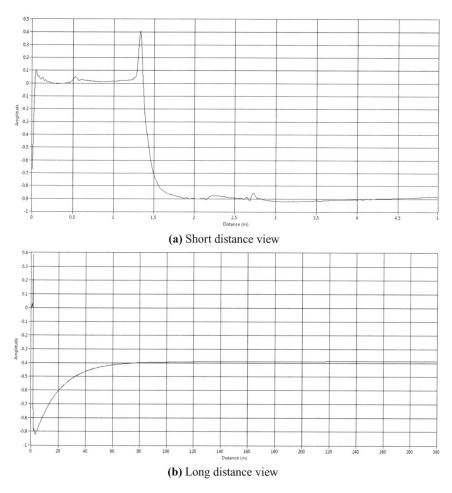

(a) Short distance view

(b) Long distance view

Fig. 4.35 Unknown load number 6

It can be noticed that the associations made with the TDR100 measuring instrument confirm the results obtained through the oscilloscope. Indeed, this laboratory experience is useful to become familiar with the TDR100 in view of the following experience.

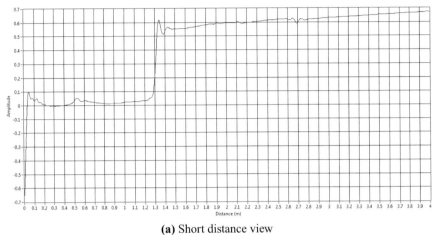

(a) Short distance view

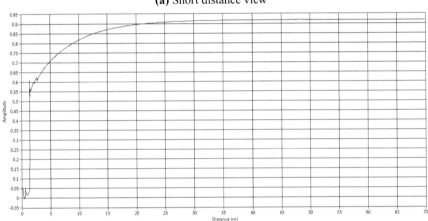

(b) Long distance view

Fig. 4.36 Unknown load number 7

Table 4.7 Case D load computation

Load type	ρ	Z_0 (Ω)	R (Ω)	D_a (m)	D_b (m)	ΔD (m)	Δt (5τ) (ns)	C (pF)
R-C series	0.62	50.00	212.45	1.34	40.01	38.67	128.89	98.22

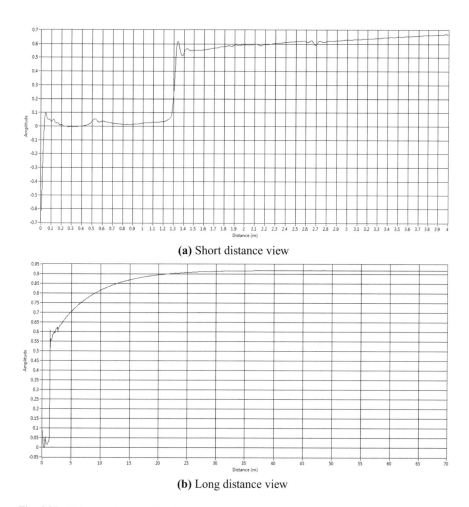

(a) Short distance view

(b) Long distance view

Fig. 4.37   Unknown load number 8

4.3.2   *Dielectric Characterization of Materials: Measurements on Water*

Equipment:
1. TDR100 reflectometer;
2. a coaxial probe for TDR measurements;
3. different types of liquids (e.g., distilled water, tap water, bottled water).

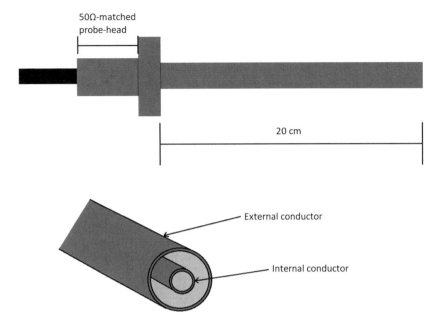

Fig. 4.38   Coaxial probe scheme

Task:
Analyze the TDR behavior of liquids with different dielectric characteristics through TDR measurements.

TDR also allows one to evaluate the dielectric characteristics of materials, in terms of relative dielectric permittivity ϵ_r. For the measurement of electrical conductivity σ (which is inversely proportional to ρ), it should ideally be measured at zero frequency; in practice, it is observed at long distances when all the transient effects have vanished. At long distances, therefore, different values of ρ can be observed, depending on the conductivity of the considered liquid [12].

TDR measurements on liquids are generally carried out through coaxial probes [13–15]. The coaxial probe (which is usually designed to have a 50 Ω load in air) is immersed in the liquid, and the liquid fills the space between an outer cylindrical conductor and an inner cylindrical conductor. As a result, the liquid acts as the filling material of a coaxial line. The electrical impedance of the probe will change depending on the dielectric characteristics of the liquid. Figure 4.38 shows a simplified sketch of the coaxial probe used for TDR measurements on three types of water.

Fig. 4.39 Commercial water
bottle

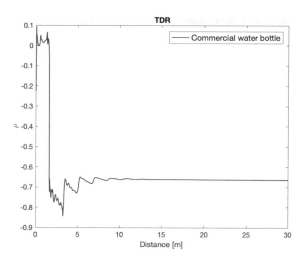

The coaxial probe used in these experiments is 20 cm long, and its end is left open-circuited.

For the TDR100 reflectometer, the output TDR waveform is express in terms of reflection coefficient as a function of the apparent (or electrical) distance. Figure 4.39 shows, the obtained TDR reflectograms, in terms of the behavior in terms of ρ versus *distance*. The observation window was 30 m long; in fact, beyond this distance, the trend was almost constant.

- **(a) Commercial water bottle.** For the commercial water bottle, it is observed that $\rho \simeq -0.74$, which is close to the value for the short circuit (Fig. 4.39).
- **(b) Distilled water.** For distilled water, it is observed that $\rho \simeq +0.66$, which is close to the value for the open circuit (Fig. 4.40).
- **(c) Tap water.** Finally, for tap water, we have $\rho \simeq -0.72$ (Fig. 4.41).

In all three cases, ϵ_r can be obtained from the apparent distance, which is approximately the same in each case.

To evaluate S_{11}, the *reflection scattering parameter*, it is first necessary to transform the TDR output into the FD. To this end, it is important to choose a suitable window such that all the multiple reflections have vanished and the signal has become almost flat (in the considered experiments, 50 m is enough). To improve accuracy of results, it is also necessary to carry out the probe calibration in order to reduce all the undesired effects (such as, spurious reflections caused at connection transitions). For the calibration, choose the same acquisition window used for the signal.

It is necessary to carry out, on the interface on which the probe begins, three TDR measurements that act as a reference; this is because they are carried by connecting three standard electric loads, that is, short circuit (S), open circuit (O), and load adapted to 50 Ω (L) (*SOL calibration*).

Fig. 4.40 Distilled water

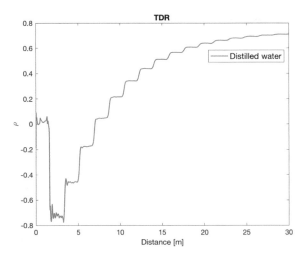

Fig. 4.41 Tap water

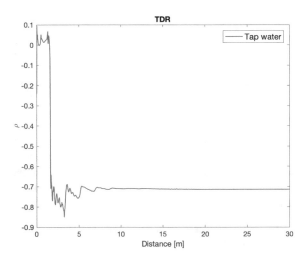

Depending on the type of probe, it is not always possible to connect standard loads for calibration. For this type of probe, a "head" is used, which has the same length as the probe, to which the loads are connected and the response of the TDR measuring system in presence of the standard electric loads can be measured. The TDR waveforms acquired for the three loads will serve for the SOL calibration.

After acquiring the TDR waveforms of the probe head with the three calibration standards, the analysis of the S_{11} parameters is performed through a very simple

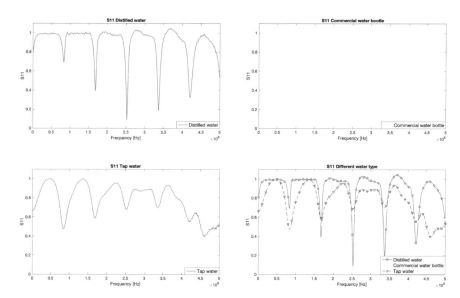

Fig. 4.42 S_{11} parameter

MATLAB script that uses the FFT algorithm. The algorithm takes into account the TDR waveforms acquired for the SOL calibration.

Once started, the script requests the following information:

- the name of the file (.txt) containing five columns, which are respectively distance (equal in all measurements), reflection coefficients of standard loads (SOL), water reflection coefficient;
- the name of the file (.txt) where the results will be stored;
- the desired frequency resolution;
- the maximum frequency of analysis desired.

In this analysis, the behavior of S_{11} up to 0.5 GHz was observed.

In Fig. 4.42, the S_{11} graphs of the different water samples are shown. These were evaluated by setting a frequency resolution of 500 Hz in the MATLAB script.

Some important observations:

- The values of S_{11} for $f = 0$ are different for the three types of water because it is related to conductivity σ.
- The S_{11} minima of the different water samples are always at the same frequencies and always equidistant (same Δf); this is because the distance between the minima is related to the dielectric permittivity ϵ_r (for the types of water analyzed, approximately 78 [10, 11]) and to the probe length L, which is known, i.e., $L = 20$ cm. The following formula is used to calculate the averaged value of the different Δf:

$$\Delta f = \frac{v_c}{2L\sqrt{\epsilon_r}} \, . \tag{4.27}$$

A simple MATLAB script allows to retrieve the values of ϵ_r; σ; and Δ_f for the considered water samples. The code is shown in Fig. 4.43.

```
% Set Probe Length in meters
L = 0.2;

% Load data from txt files
data1=load('S11_AcquaDistillata.txt');
data2=load('S11_AcquaBottiglia.txt');
data3=load('S11_AcquaRubinetto.txt');

% Extract frequency and module vectors from data
frequency1=data1(:,1);
mod1=data1(:,2);
frequency2=data2(:,1);
mod2=data2(:,2);
frequency3=data3(:,1);
mod3=data3(:,2);

% Relative minimum points search
TF1 = islocalmin(mod1,'MinSeparation',80*10^6,'SamplePoints',frequency1);
TF2 = islocalmin(mod2,'MinSeparation',80*10^6,'SamplePoints',frequency2);
TF3 = islocalmin(mod3,'MinSeparation',80*10^6,'SamplePoints',frequency3);

% Extract frequency values of minimum points
minFreq1=frequency1(TF1);
minFreq2=frequency2(TF2);
minFreq3=frequency3(TF3);

% Calculate 3 sample of frequency range between two peaks
deltaf1=ones(3,1);
deltaf2=ones(3,1);
deltaf3=ones(3,1);
for i = 1:3
    deltaf1(i,:)=minFreq1(i+1,:)-minFreq1(i,:);
    deltaf2(i,:)=minFreq2(i+1,:)-minFreq2(i,:);
    deltaf3(i,:)=minFreq3(i+1,:)-minFreq3(i,:);
end

% Calculate average of the 3 samples
deltafm1=median(deltaf1);
deltafm2=median(deltaf2);
deltafm3=median(deltaf3);

% Calculate the electrical permittivity with formula: eps=c^2/(?f^2*4*L^2)
epsilon_distilled=physconst('LightSpeed')^2/(deltafm1^2*4*L^2);
epsilon_bottle=physconst('LightSpeed')^2/(deltafm2^2*4*L^2);
epsilon_tap=physconst('LightSpeed')^2/(deltafm3^2*4*L^2);

% Print ?f, electrical permittivity and S11 in DC
disp("Distilled water")
disp("Delta-f " + deltafm1/10^6 + " MHz")
disp("Epsilon " + epsilon_distilled)
disp("S11-DC  " + -mod1(1,1))
disp(" ")
disp("Bottle water")
disp("Delta-f " + deltafm2/10^6 + " MHz")
disp("Epsilon " + epsilon_bottle)
disp("S11-DC  " + mod2(1,1))
disp(" ")
disp("Tap water")
disp("Delta-f " + deltafm3/10^6 + " MHz")
disp("Epsilon " + epsilon_tap)
disp("S11-DC  " + mod3(1,1))
```

```
>> TDRwater
Distilled water
Delta-f 84.4635 MHz
Epsilon 78.7378
S11-DC  -0.80928

Bottle water
Delta-f 84.954 MHz
Epsilon 77.8312
S11-DC  0.62529

Tap water
Delta-f 83.701 MHz
Epsilon 80.1789
S11-DC  0.66285
>>
```

Fig. 4.43 MATLAB script for Δf, ϵ_r, and σ computations

References

1. Cataldo A, Cannazza G, De Benedetto E, Giaquinto N (2012) Experimental validation of a TDR-based system for measuring leak distances in buried metal pipes. Prog Electromagn Res 132:71–90
2. Cataldo A, De Benedetto E, Cannazza G (2015) Hydration monitoring and moisture control of cement-based samples through embedded wire-like sensing elements. IEEE Sens J 15(2):Art no 6912927, 1208–1215
3. Cataldo A, De Benedetto E, Cannazza G, Masciullo A, Giaquinto N, D'Aucelli GM, Costantino N, De Leo A, Miraglia M (2017) Recent advances in the TDR-based leak detection system for pipeline inspection. Meas: J Int Meas Confed 98:347–354
4. Cataldo A, De Benedetto E, Cannazza G, (2011) Broadband reflectometry for enhanced diagnostics and monitoring applications. Springer, Berlin
5. Time domain reflectometry theory (2006) Agilent Application Note 1304–2. Palo Alto, CA
6. Network analyzer basics (2000) Agilent Technologies, USA. http://materias.fi.uba.ar/6644/info/anredes/basico/Network%20Analyzer%20Basics.pdf
7. High precision time domain reflectometry (2003) Agilent Application Nite 1304–7, USA
8. Friedrich N (2013) Vector network analysis: a quick rundown on the basics. https://www.mwrf.com/test-amp-measurement/vector-network-analysis-quick-rundown-basics(Nov04,2013)
9. TDR100 instruction manual - revision 5/15. Logan, UT (2010) https://s.campbellsci.com/documents/au/manuals/tdr100.pdf
10. Akhadov YY (1980) Dielectric properties of binary solutions. Pergamon, Oxford
11. Buckley F, Maryott AA (1958) Tables of dielectric dispersion data for pure liquids and dilute solutions. National Bureau of Standards Circular 589, Washington, DC
12. Castiglione P, Shouse PJ (2003) The effect of ohmic losses on time-domain reflectometry measurements of electrical conductivity. Soil Sci Soc Am J 67:414–424
13. Cataldo A, Catarinucci L, Tarricone L, Attivissimo F, Trotta A (2007) A frequency-domain method for extending TDR performance in quality determination of fluids. Meas Sci Technol 18(3):675–688
14. Cataldo A, Tarricone L, Attivissimo F, Trotta A (2007) A TDR method for real-time monitoring of liquids. IEEE Trans Instr Meas 56(8):1616–1625
15. Cataldo A, Vallone M, Tarricone L, Attivissimo F (2008) An evaluation of performance limits in continuous TDR monitoring of permittivity and levels of liquid materials. Measurement 41(7):719–730

Correction to: Basic Theory and Laboratory Experiments in Measurement and Instrumentation

Correction to:
A. Cataldo et al., *Basic Theory and Laboratory*
Experiments in Measurement and Instrumentation,
Lecture Notes in Electrical Engineering 663,
https://doi.org/10.1007/978-3-030-46740-1

The original version of the book was inadvertently published with an incorrect spelling of the first name of the author Alessandro De Monte. The first name was corrected from Alessando to Alessandro. The corrected book has been updated with the changes.

The updated version of the book can be found at
https://doi.org/10.1007/978-3-030-46740-1

Appendix
PCB Scheme

This appendix contains the schematics of the printed circuit board that was used for the laboratory experiences. Figure A.1 shows a picture of the fabricated PCB. The general schematic is shown in Fig. A.2. The diagram includes all the circuits considered in the laboratory experiences. The use of jumpers to short-circuit specific pins makes it possible to obtain the different circuital configurations (e.g., RC filter and CR filter). The position of the jumpers is indicated with thick green lines in the diagrams (Figs. A.3, A.4, A.5, A.6, A.7, A.8, A.9, A.10).

© Springer Nature Switzerland AG 2020   181
A. Cataldo et al., *Basic Theory and Laboratory Experiments in Measurement and Instrumentation*, Lecture Notes in Electrical Engineering 663,
https://doi.org/10.1007/978-3-030-46740-1

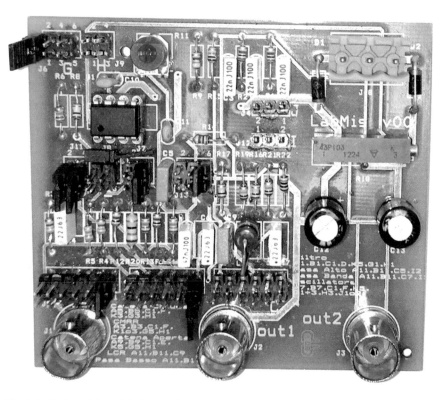

Fig. A.1 Picture of the fabricated PCB

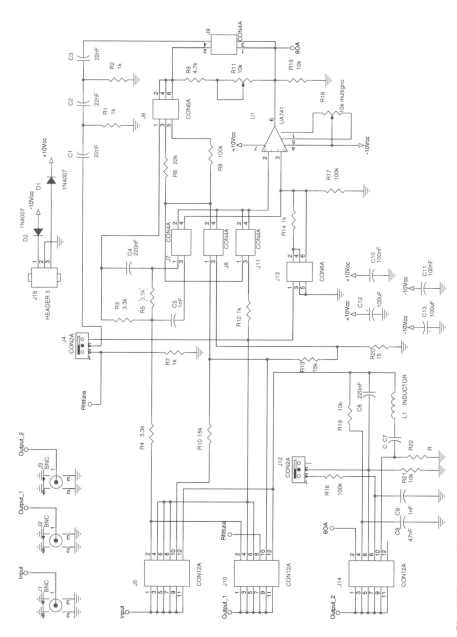

Fig. A.2 Circuit diagram of the PCBs used

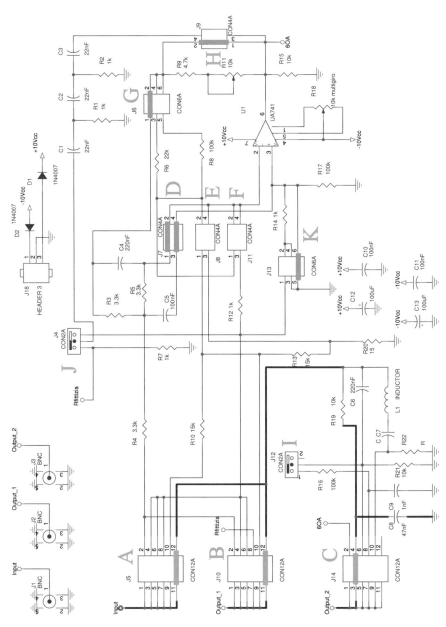

Fig. A.3 Circuit diagram of the PCBs used: configuration for the RC filter

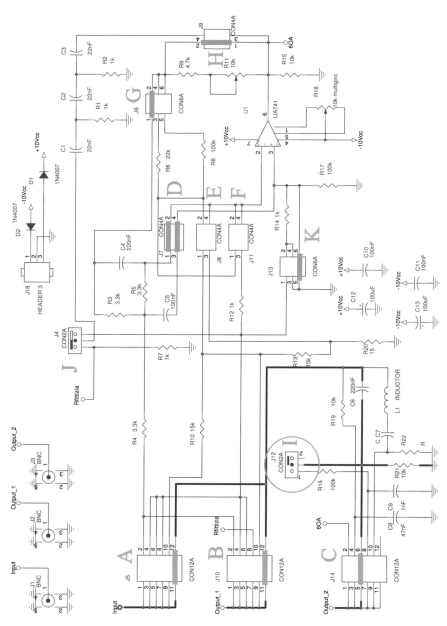

Fig. A.4 Circuit diagram of the PCBs used: configuration for the CR filter

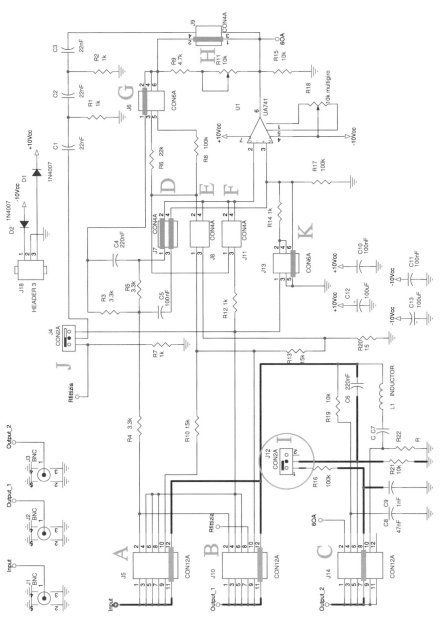

Fig. A.5 Circuit diagram of the PCB: configuration for the CR-RC filter

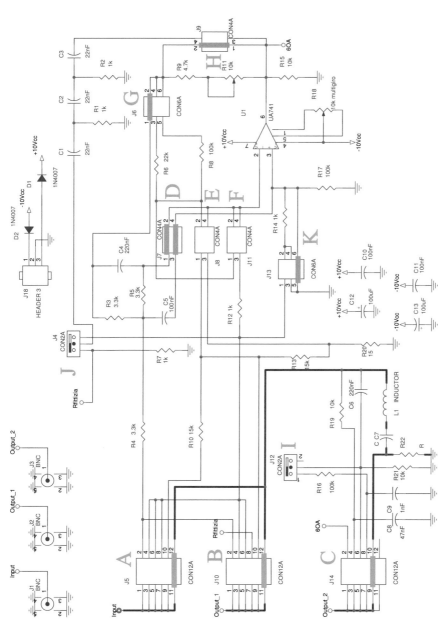

Fig. A.6 Circuit diagram of the PCBs used: configuration for the LCR filter

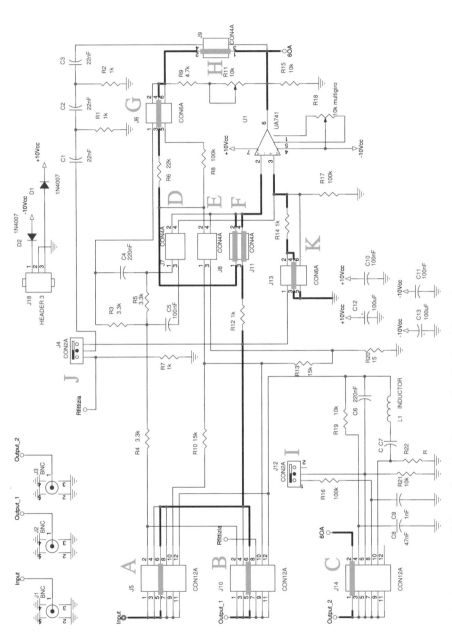

Fig. A.7 Circuit diagram of the PCB: configuration for evaluating the closed-loop gain

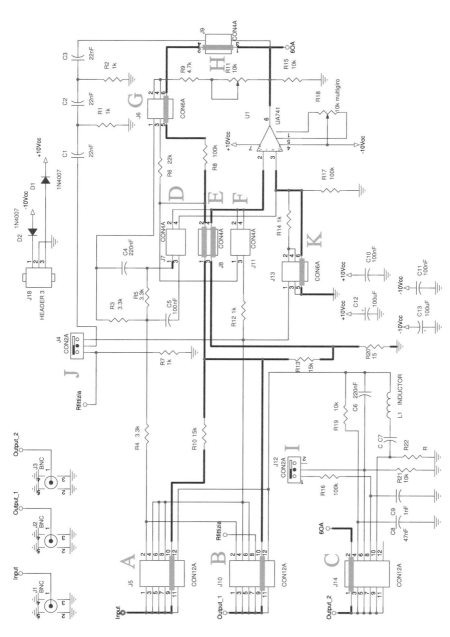

Fig. A.8 Circuit diagram of the PCB: configuration for evaluating the open-loop gain

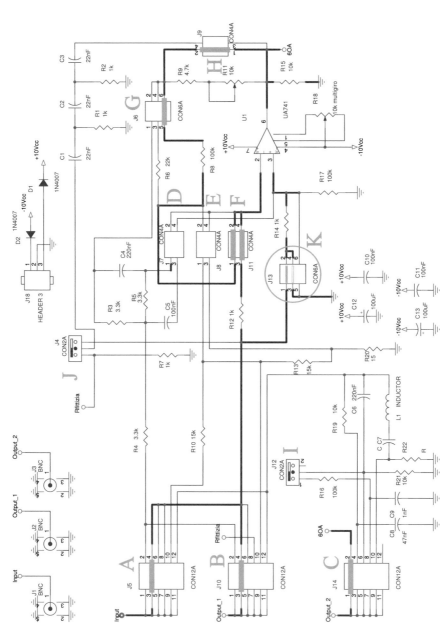

Fig. A.9 Circuit diagram of the PCB: configuration for evaluating CMRR

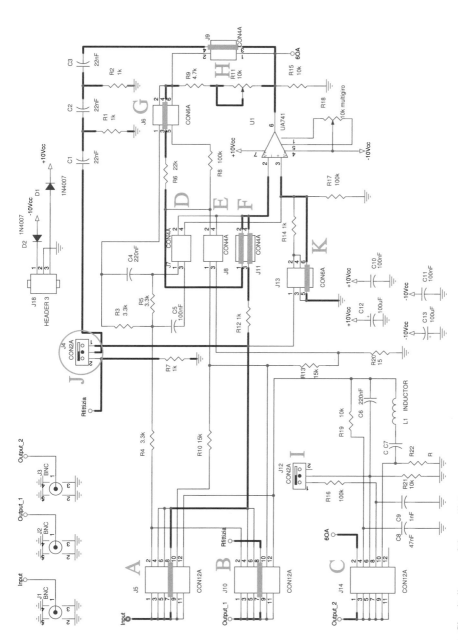

Fig. A.10 Circuit diagram of the oscillator